WESTHOLLOW RESEARCH CENTER LIBRARY
SHELL DEVELOPMENT COMPANY
P. O. BOX 1378
HOUSTON, TX 77001. U.S.A.

DEVELOPMENTS IN POLYMER DEGRADATION—2

THE DEVELOPMENTS SERIES

Developments in many fields of science and technology occur at such a pace that frequently there is a long delay before information about them becomes available and usually it is inconveniently scattered among several journals.

Developments Series books overcome these disadvantages by bringing together within one cover papers dealing with the latest trends and developments in a specific field of study and publishing them within *six months* of their being written.

Many subjects are covered by the series, including food science and technology, polymer science, civil and public health engineering, pressure vessels, composite materials, concrete, building science, petroleum technology, geology, etc.

Information on other titles in the series will gladly be sent on application to the publishers.

FEB 11 198
FEB 11 1980

DEVELOPMENTS IN POLYMER DEGRADATION—2

Edited by

N. GRASSIE
Chemistry Department, The University, Glasgow, UK

WESTHOLLOW RESEARCH CENTER LIBRARY
SHELL DEVELOPMENT COMPANY
P. O. BOX 1378
HOUSTON, TX 77001. U.S.A.

APPLIED SCIENCE PUBLISHERS
LONDON

APPLIED SCIENCE PUBLISHERS LTD
RIPPLE ROAD, BARKING, ESSEX, ENGLAND

British Library Cataloguing in Publication Data

Developments in polymer degradation.—
(Developments series).
2
1. Polymers and polymerization—Deterioration
I. Grassie, Norman II. Series
668.4'1 TP1122

ISBN 0-85334-854-5

WITH 15 TABLES AND 128 ILLUSTRATIONS

© APPLIED SCIENCE PUBLISHERS LTD 1979

All rights reserved. No part of this publication may be reproduced, stored in a retrieval system, or transmitted in any form or by any means, electronic, mechanical, photocopying, recording, or otherwise, without the prior written permission of the publishers, Applied Science Publishers Ltd, Ripple Road, Barking, Essex, England

Printed in Great Britain by Galliard (Printers) Ltd, Great Yarmouth

PREFACE

The authors of the second volume of the series are all once again very actively engaged in research in the field of polymer degradation and have undertaken to describe and discuss some of their recent contributions in the context of the general development of this area of polymer science.

Although it is to be anticipated that a comprehensive picture of the current state of knowledge of polymer degradation will emerge as further volumes are published, it is clearly not possible to present a broad view of the whole subject in the limited number of contributions contained in this present publication. Nevertheless, two themes of very wide current interest will be evident. Thus, in the first four chapters the emphasis is placed strongly upon the application of experimental techniques, some already well established in chemistry generally and others specifically designed for the solution of problems in polymer degradation. On the other hand, the general theme of the remaining chapters relates to the oxidation of polymers, both thermal and photo which are of such vital importance in the exploitation of commercial materials.

In the first chapter Dr Kiran and Professor Gillham describe a complex system in which a programmable pyrolyser, a thermal conductivity detector, a mass chromatograph, a gas chromatograph, a fast-scan vapour phase infrared spectrophotometer and a computer are combined in such a way as to provide simultaneously a large amount of experimental data. They demonstrate its application in particular to the analysis of the thermal history and decomposition phenomena which occur in polyolefins, polyolefinsulphones, polymethacrylates, polystyrenes and polybutadienes. This is followed, in Chapter 2, by a description by Dr Schnabel of the investigation by the light scattering technique of such diverse phenomena as chain scission and cross-linking, radical lifetimes and diffusional processes induced in some synthetic and biological polymers by pulsed high-energy electron and ultra-violet light irradiation. In Chapter 3, Dr Lüderwald shows how the complex series of reactions which occurs during the thermal degradation of polyesters may be unravelled by mass spectrometric analysis after pyrolysis in the ion source. Professor Sohma then demonstrates, in Chapter 4, the power of electron spin resonance spectroscopy as an analytical tool. He reports upon the nature of the radicals produced by mechanical fracture of solid polymers and by ultrasonic irradiation of polymer solutions. He discusses their thermal decay and photoconversion and a variety of related topics.

Chapters 5 and 6 are devoted to photooxidation phenomena. Drs Allen and McKellar demonstrate that fluorescent and phosphorescent impurities are present in a number of commercial polymers including polyolefins, polyamides, polyurethanes and poly(ethersulphone)s. The luminescent species have been identified and their rôle in the photooxidation process clarified. Dr Arnaud and Professor Lemaire report upon their investigations into the photooxidation of polyethylene and polypropylene catalysed by titanium and zinc oxides. In the final chapter Professor Tüdős and Drs Kelen and Nagy compare the thermal and thermo-oxidative reactions which occur in poly(vinyl-chloride), demonstrating in particular the rôle of polyene sequences in the partially degraded polymer.

It is hoped that these discussions, like those contained in the first volume of this series, will be of interest to advanced students, research workers and technologists who are concerned with the reactions, properties and applications of polymers.

I take this opportunity to thank the authors and publishers for the efficiency of their collaboration in the production of this volume.

N. GRASSIE

CONTENTS

LIST OF CONTRIBUTORS

NORMAN S. ALLEN

*Department of Chemistry, John Dalton Faculty of Technology, Manchester Polytechnic, Chester Street, Manchester M*1 *5GD, UK.*

RENÉ ARNAUD

Laboratoire de Photochimie, UER Sciences Exactes et Naturelles, Université de Clermont-Ferrand, 63170 *Aubière, France.*

JOHN K. GILLHAM

Polymer Materials Program, Department of Chemical Engineering, Princeton University, Princeton, New Jersey 08540, *USA.*

T. KELEN

*Central Research Institute for Chemistry of the Hungarian Academy of Sciences, H-*1025 *Budapest, Hungary.*

ERDOĞAN KIRAN

SEKA (Türkiye Selüloz ve Kağit Fabrikalari Isletmesi), Central Research Laboratory, Izmit, Kocaeli, Turkey.

JACQUES LEMAIRE

Laboratoire de Photochimie, UER Sciences Exactes et Naturelles, Université de Clermont-Ferrand, 63170 *Aubière, France.*

INGO LÜDERWALD

Institute of Organic Chemistry, University of Mainz, Johann-Joachim-Becher-Weg 18–20, *D*-6500 *Mainz, West Germany.*

JOHN F. MCKELLAR

*Department of Chemistry and Applied Chemistry, University of Salford, Salford M*5 4*WT, Lancashire, UK.*

T. T. NAGY

Central Research Institute for Chemistry of the Hungarian Academy of Sciences, H-1025 *Budapest, Hungary.*

WOLFRAM SCHNABEL

Hahn-Meitner-Institut für Kernforschung Berlin GmbH, Bereich Strahlenchemie, Glienicker Strasse 100, *D*-100 *Berlin* 39, *West Germany.*

JUNKICHI SOHMA

Faculty of Engineering, Hokkaido University, Sapporo 060, *Japan.*

F. TÜDŐS

Central Research Institute for Chemistry of the Hungarian Academy of Sciences, H-1025 *Budapest, Hungary.*

Chapter 1

PYROLYSIS–MOLECULAR WEIGHT CHROMATOGRAPHY–VAPOUR PHASE INFRARED SPECTROPHOTOMETRY: AN ON-LINE SYSTEM FOR ANALYSIS OF POLYMERS

ERDOĞAN KIRAN

SEKA (Türkiye Selüloz ve Kağit Fabrikalari Isletmesi), Central Research Laboratory, Turkey

and

JOHN K. GILLHAM

Princeton University, USA

SUMMARY

An instrumental system consisting of a combination of a programmable pyrolyser, a thermal conductivity detector, a mass chromatograph, a gas chromatograph, a fast-scan vapour phase infrared spectrophotometer and a computer is discussed and reviewed with examples which show the utility of the system in the analysis of decomposition phenomena in polyolefins, polyolefinsulphones, polymethacrylates, polystyrenes, and polybutadienes.

The system records the thermal history before and during pyrolysis of the sample and provides chromatographic retention times, infrared spectra, mass numbers and relative amounts of the volatile products of pyrolysis. Identification of the constituents of the effluents is thus facilitated and mechanisms for decomposition become easier to propose and to verify.

1. INTRODUCTION

Pyrolytic decomposition followed by analysis of the volatile products of pyrolysis is a powerful method for characterisation of polymeric materials. The scope and applicability of this approach are determined by the type of pyrolyser and the separation and identification techniques that are utilised.

The on-line system developed at Princeton University comprises a programmable pyrolyser, a thermal conductivity cell, a trap, a mass chromatograph, a conventional gas chromatograph, a vapour phase infrared spectrophotometer, and a digital computer.[1–3] The system has proved powerful for analysis of polymers as has been shown by results obtained with polyolefins,[1,4–7] polyolefinsulphones,[1,8] polymethacrylates,[1] polystyrenes[1,3,8,9] and polybutadienes.[10] This report provides a brief description of the system and reviews results obtained with these polymers.

2. THE PYROLYSIS LABORATORY

A schematic diagram of the system is shown in Fig. 1. A critical discussion of its various parts has been presented elsewhere.[2] Briefly, the system consists of a programmable pyrolyser, a thermal conductivity cell (TC), a trap, a mass chromatograph (MC), a conventional gas chromatograph (GC), a vapour phase infrared spectrophotometer (Vapour Phase IR) and a digital computer (IBM System 7).

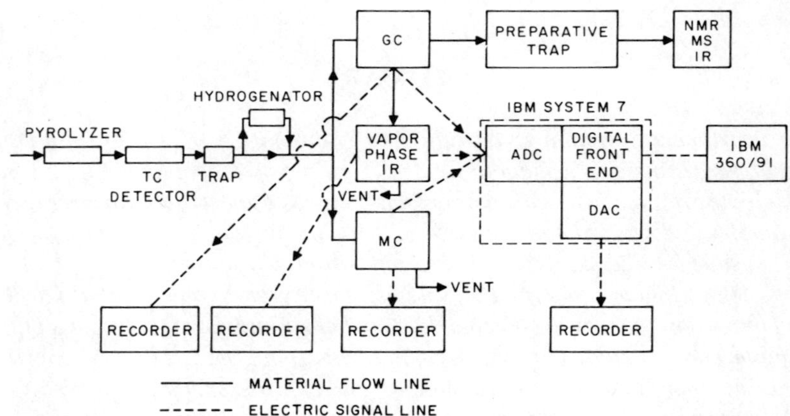

FIG. 1. Schematic diagram of the laboratory for pyrolytic studies.

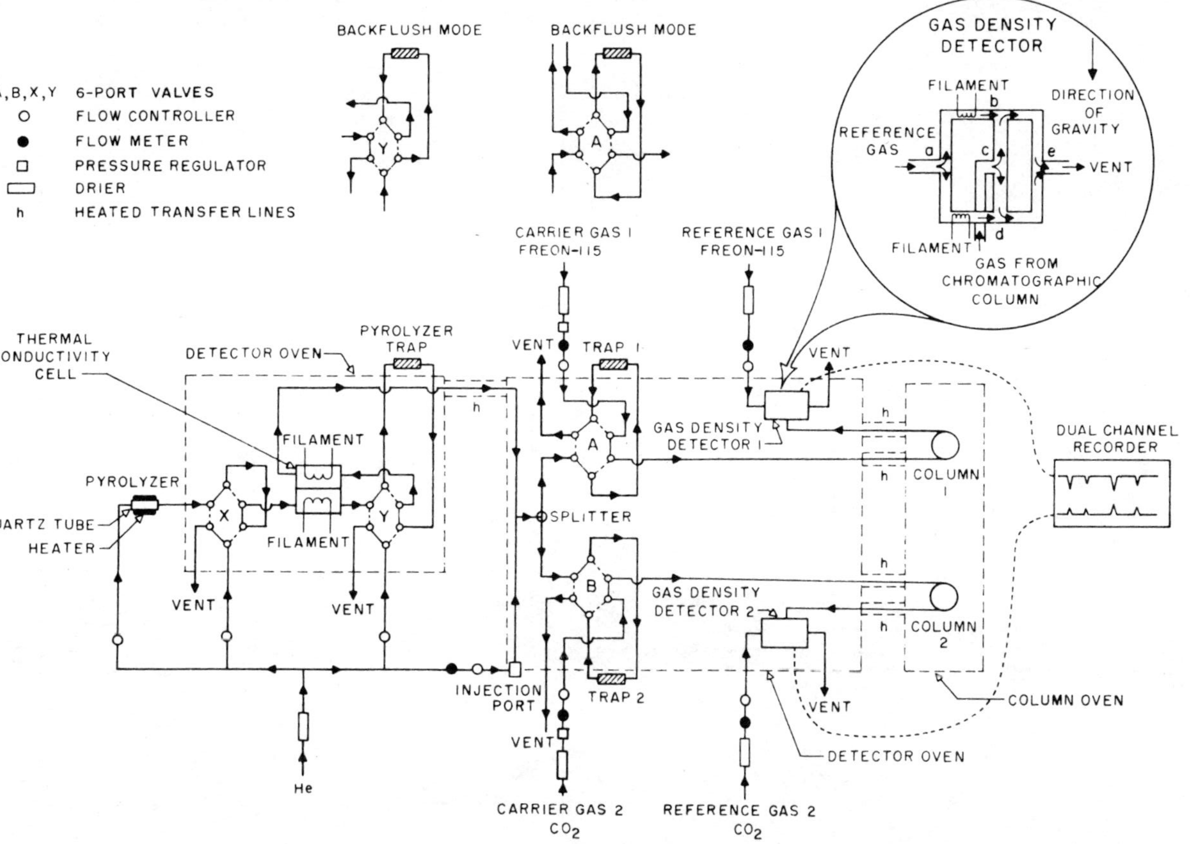

FIG. 2. Coupling of the pyrolyser with the thermal conductivity detector, the trap, and the mass chromatograph.

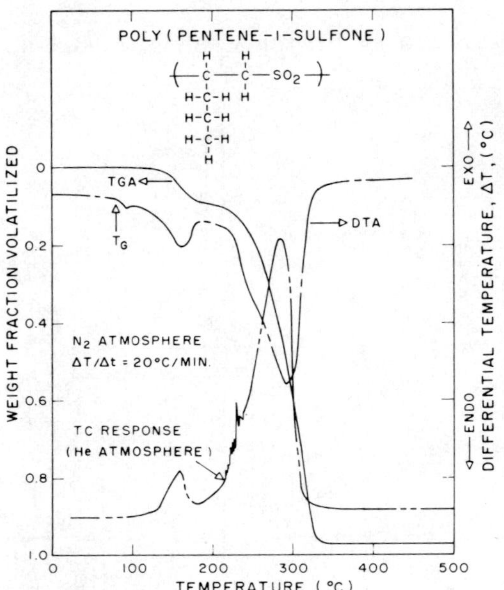

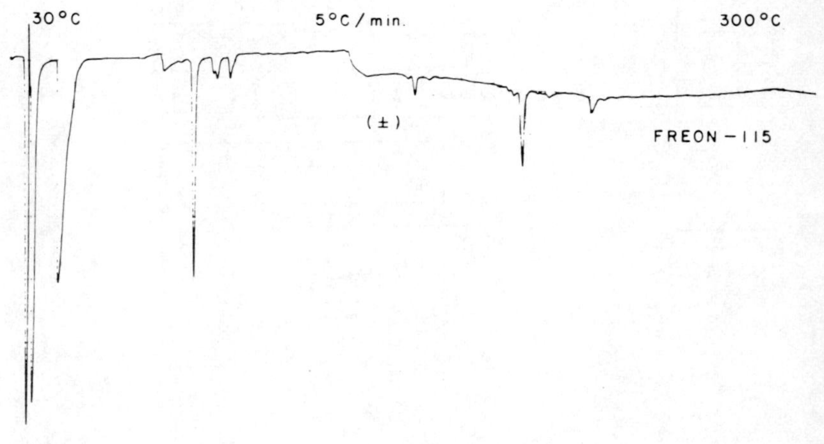

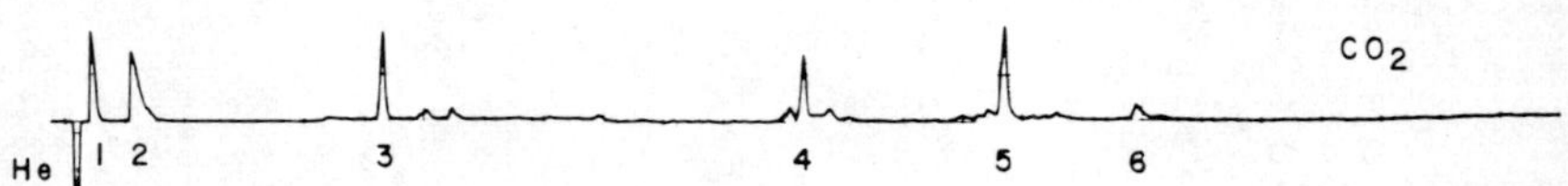

FIG. 3. Top: Thermal history before and during pyrolysis of poly(pentene-1-sulphone) ($[\eta] = 1{\cdot}49$ dl/g). Below: Mass chromatogram of the volatile pyrolysis products of poly(pentene-1-sulphone). Peak attenuations were $\times 256$ for peaks 1 and 2, and $\times 16$ thereafter. The columns (SE-30) were program-heated at 5 °C/min from 30° to 300 °C.

Figure 2 shows details of the coupling of the pyrolyser with the thermal conductivity detector, the trap, and the mass chromatograph.

The pyrolyser consists of a quartz tube and an external heater which provides flexibility for either program-heating or flash-heating.

The thermal conductivity cell monitors the formation of the volatile products of pyrolysis as a function of temperature and time. It provides immediate information about the onset and the progress of the thermal events that result in the formation of volatile products. The response curves contain features that are characteristic of the material under investigation.

The combination of the pyrolyser with the thermal conductivity detector is complementary to a thermogravimetric analyser and/or differential thermal analyser. As an example, the thermal history before and during pyrolysis of poly(pentene-1-sulphone) as recorded with the thermal conductivity cell, a thermogravimetric analyser (TGA) and a differential thermal analyser (DTA) is shown in Fig. 3.

The trap is an important part of the interface between the pyrolysis assembly and the chromatographic assembly. It is a short stainless steel column packed with Porapak Q (Waters Associates) and has a geometry which permits a fraction of it to be placed in a Dewar flask for subambient cooling. The function of the trap is to collect the fragments of pyrolysis that are formed in the temperature range of interest (i.e. selective trapping). The products are released from the trap by rapid heating, and introduced as a slug to the chromatographic columns for separation. The valving system permits monitoring of the release of the products from the trap with the same thermal conductivity cell used to monitor the formation of volatile products; the response is recorded as a sharp peak, as shown, for example in Fig. 4.

The mass chromatograph is a special type of gas chromatograph which directly provides mass numbers of the resolved components of a mixture by means of a pair of gas density detectors.[11] The instrument consists of two independent gas chromatographs which use a common injection port (see Fig. 2). When a sample is introduced to the instrument it is split into two, approximately equal, fractions (splitter, Fig. 2) and each fraction is carried to a trap which is similar to the trap associated with the pyrolyser. Then, through a valve arrangement, different carrier gases are introduced to the two traps; the samples, released by rapid heating of the traps, are carried over to two matched chromatographic columns for separation. The constituents eluted from the chromatographic columns are detected by gas density detectors. The recorder output from the mass chromatograph displays two sets of peaks corresponding to the responses from the two gas

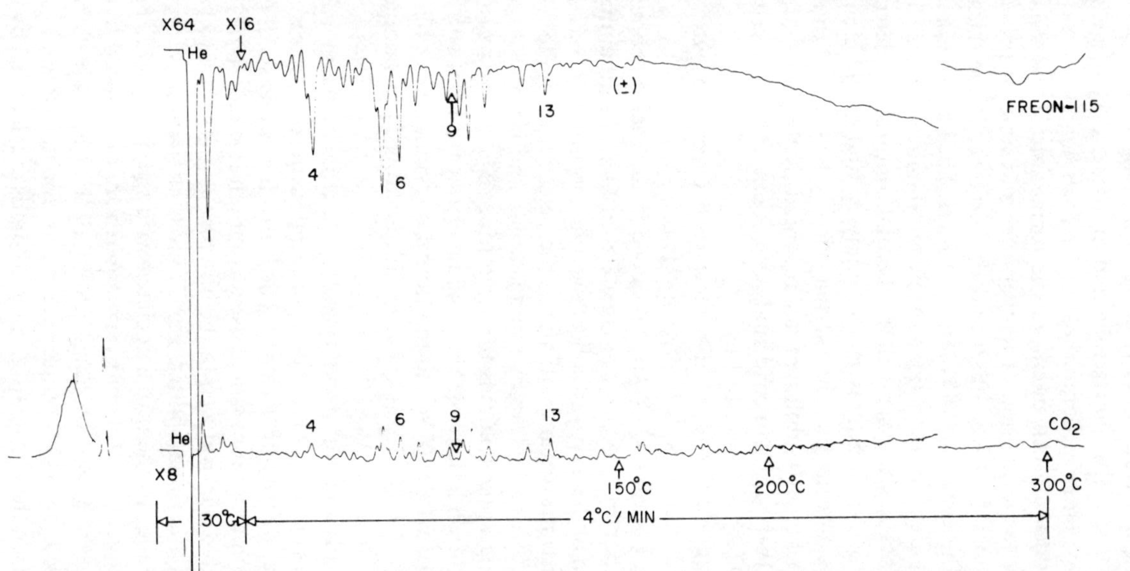

FIG. 4. Mass chromatogram of the volatile pyrolysis products of syndiotactic 1,2-polybutadiene (1,2 structure 83–90 %, $\bar{M}_w = 177\,000$, $\bar{M}_n = 107\,000$) formed during 100 % weight loss. Peak attenuations for the Freon-115 channel were × 64 for peak 1, and × 16 thereafter. Those for the CO_2 channel were × 8 for all peaks. The columns (Dexsil-300) were program-heated at 4 °C/min from 30 ° to 300 °C. The broad peak at the extreme left shows the thermal conductivity cell response during pyrolysis; the subsequent sharp peak shows the response during flash release of the products from the pyrolyser trap.

chromatographic systems with different carrier gases for the same constituents of the mixture (see Figs. 3 and 4).

Theoretical aspects and operational principles of the mass chromatograph and the gas density detector are presented elsewhere.[1,11] The ratio of the responses is related to the molecular weight (M_x) of a constituent through

$$M_x = \frac{\left(\frac{A_1}{A_2}\right) K M_{C_2} - M_{C_1}}{\left(\frac{A_1}{A_2}\right) K - 1}$$

where M_{C_1} and M_{C_2} are the molecular weights of the carrier gases 1 and 2, A_1 and A_2 are the chromatographic peak areas for the constituent, and K is the instrument constant determined from analysis of samples of known molecular weights from:

$$K = \left(\frac{M_x - M_{C_1}}{M_x - M_{C_2}}\right) \bigg/ \left(\frac{A_1}{A_2}\right)$$

The response of the gas density detector is proportional to $(M_x - M_C)/M_x$ and therefore, for a given carrier gas of molecular weight M_C, the response tends to a constant as M_x increases; the lower the molecular weight of the carrier gas, the faster is this tendency. In order to cover a wide range of molecular weights, the two carrier gases are chosen to be of highly different molecular weights. The gases used in these studies were CO_2 and ClC_2F_5 (Freon-115). Use of CO_2 and SF_6 is reported[12] in a study of oxidative degradation of isotactic poly(1-pentene) using a combination of a thermal conductivity detector and a mass chromatograph.

Since the molecular weights of the carrier gases are known and since the response ratios (A_1/A_2) can be measured from the chromatographic output, calculation of the molecular weights of the constituents becomes a simple process after the instrument constant is evaluated. In principle, the instrument constant K can be evaluated by injecting just one solute (of known molecular weight) into the mass chromatograph and measuring the response ratio from the output. However, the accuracy of measuring the response ratio varies with molecular weight and consequently K must be evaluated by using more than one compound if it is to be applicable to a wide range of molecular weights. Otherwise instantaneous K values applicable in small ranges are used.

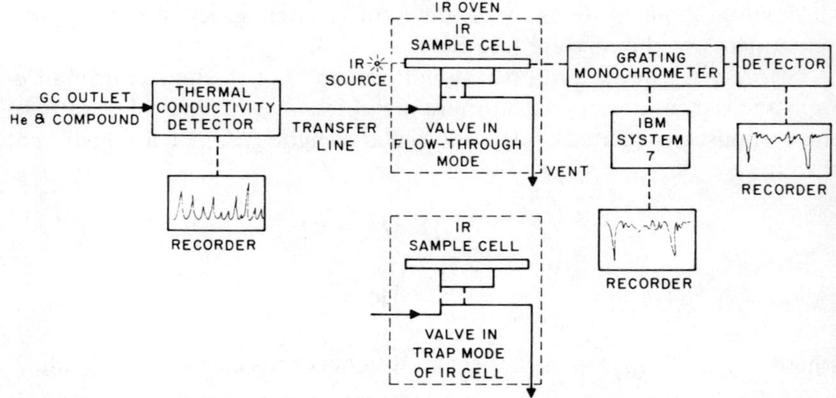

FIG. 5. Schematic diagram of the coupling of the gas chromatograph with the fast-scan vapour phase infrared spectrophotometer and the computer.

In addition to providing mass numbers of resolved constituents the mass chromatograph can be used to provide estimates of the weight amounts of each constituent (w_i). This is performed by comparing the areas on the CO_2 channel output of the mass chromatograph with those of peaks obtained under the same chromatographic conditions from a synthetic mixture containing known amounts of n-hydrocarbons.[3] For compounds of very similar molecular weight and with a similar retention time, responses of the gas density balance detector are directly proportional to their amounts (see eqn. 71 of reference 11). The relative amount of a component as a weight percentage (P_{wt}) of those measured can be calculated using $P_{wt} = (w_i/\Sigma w_i) \times 100$.

A fast-scan vapour phase infrared spectrophotometer is suitable for taking 'on-the-fly' spectra of constituents eluting from chromatographic columns. The instrument is capable of scanning from 2·5 to 15 μm (4000 to 670/cm) in either 6 or 30 s. The spectrophotometer can be coupled with either the mass chromatograph or the conventional gas chromatograph. When coupled with the mass chromatograph sensitivity is reduced due to the fact that column effluents are greatly diluted by the reference gas entering the gas density detector. Furthermore the infrared (IR) spectra of the carrier gases, CO_2 and Freon-115, are complex and can mask spectra of the constituents.[2] Therefore coupling of an infrared spectrophotometer with a conventional gas chromatograph is preferred. A schematic diagram of the coupling of the gas chromatograph with the infrared spectrophotometer and the computer is shown in Fig. 5.

The gas chromatograph used is a computer-compatible, automated research instrument equipped with thermal conductivity and flame ionisation detectors.

The computer facility is a flexible mini-computer capable of pre-processing data for analysis by a much larger computational system (to which it is connected by telephone). As a constituent elutes from the chromatographic column it may be trapped in the IR cell and a set number of IR spectra taken. These spectra are computer-averaged and the background spectrum is subtracted in order to increase the signal-to-noise ratio.

3. THERMAL DECOMPOSITION OF POLYMERS

Thermal decomposition of various polymers has been studied by the on-line system just described. Polymers studied include: (i) polyolefins;[1,4–7] (ii) polyolefinsulphones;[1,8] (iii) polymethacrylates;[1] (iv) polystyrenes[1,3,8,9] and (v) polybutadienes.[10] Comparative discussions with literature reports on the same polymers using different techniques have been presented in references 1–11.

3.1. Polyolefins

Polyethylene, polypropylene, and polyisobutylene samples were pyrolysed in flowing helium from room temperature to 600 °C, rising by 20 °C/min. Products volatilising in the temperature range 300–600 °C were selected for trapping and analysis.

The output from the mass chromatograph for low density polyethylene is shown in Fig. 6. The regularly spaced doublet peaks were shown to correspond to alkenes and alkanes by calculating the molecular weights and comparing the retention times with hydrocarbons.[1,7] Molecular weights calculated from the mass chromatographic output for some of the assigned structures are given in Table 1. The formation of the hydrocarbon constituents seen in Fig. 6 and the observed relative abundance of C_6, C_{10}, C_{14} and C_{18} alkenes and C_3, C_7, C_{11} and C_{15} alkanes were considered to follow from an intramolecular radical transfer process which involves the 5th carbon atom of the primary macroradical through a pseudo six-membered ring intermediate. The presence of ethylene is not seen in Fig. 6 since the column conditions were not suitable for its analysis.

Main-chain scission in polyethylene results initially in primary radicals which can undergo either depolymerisation or intra- and/or intermolecular

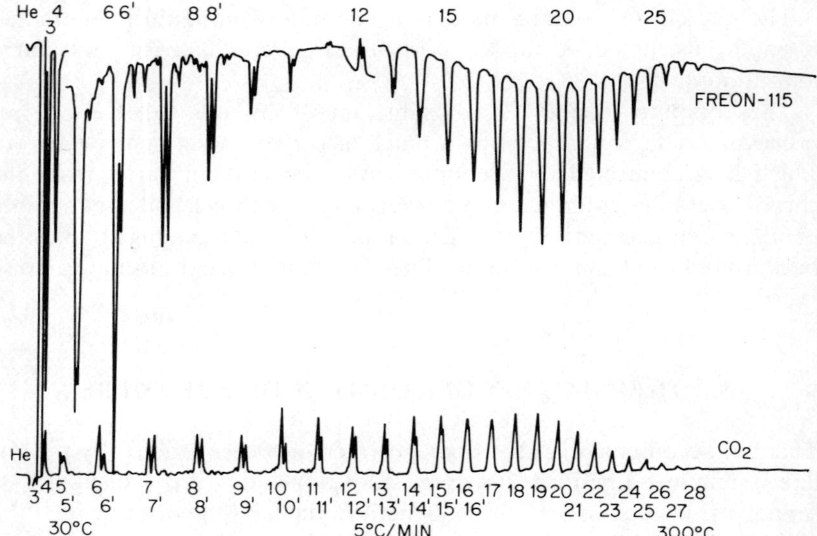

FIG. 6. Mass chromatogram of the volatile pyrolysis products of low density polyethylene ($\bar{M}_w = 140\,000$, $\bar{M}_n = 16\,000$). Peak attenuations were $\times 8$, except for peaks 3 and 4 in the Freon-115 channel for which the attenuations were $\times 64$. Columns (SE-30) were program-heated at 5 °C/min from 30 to 300 °C.

radical transfer processes. The relative extents of the competing reactions determines the degradation pattern.

Intramolecular radical transfer reactions in primary radicals (**I**) lead to the formation of secondary radicals (**II**). These can undergo β-scission to the left of the radical centre and produce alkenes (**III**).

$$\sim CH_2CH_2\overset{\overset{\displaystyle H}{|}}{C}H(CH_2)_nCH_2\cdot \rightarrow \sim CH_2CH_2\dot{C}H(CH_2)_nCH_3$$

(**I**) (**II**)

$$\mathbf{II} \rightarrow \sim CH_2\cdot + CH_2{=}CH(CH_2)_nCH_3$$

(**III**)

If the β-scission is to the right of the radical centre, short-chain primary radicals (**IV**) form

$$\mathbf{II} \rightarrow \sim CH_2CH_2CH{=}CH_2 + \cdot CH_2(CH_2)_{n-2}CH_3$$

(**IV**)

TABLE 1

CALCULATED MOLECULAR WEIGHTS AND ASSIGNED STRUCTURES FOR SOME OF THE CHARACTERISTIC PEAKS IN THE PYROGRAMS DISCUSSED

Figure number	*Peak number*	*Assigned structure*	*Molecular weight*	
			Theoretical	*Calculated*
6	6	Hexene	84	85·1
	6′	Hexane	86	85·5
	10	Decene	140	141·6
	10′	Decane	142	142·9
	14	Tetradecene	196	195·4
	20	Eicosene + (eicosane)	(283)	287·4
7	3	Pentane	72	77
	5	2-Methylpentane	86	86
	9	2,4-Dimethyl-1-heptene	126	124·7
	11	2,4,6-Trimethylnonane	170	171
	12	2,4,6,8-Tetramethyl-1-hendecene	210	210·7
	16	2,4,6,8,10,12-Hexamethyl-1-tridecene	266	270·4
8	6	2,2,4-Trimethylpentane	114	112·2
	12	2,4,4,6,6-Pentamethyl-1-heptene	168	170·5
	16	2,4,4,6,6,8,8-Heptamethyl-1-nonene	224	227
	20	2,4,4,6,6,8,8,10,10-Nonamethyl-1-hendecene	280	277·2
	24	2,4,4,6,6,8,8,10,10,12,12-Undecamethyl-1-tridecene	336	325
11	1	Methylmethacrylate	100	100·3
13	1	Isobutylene	56	60·1
	2	t-Butylmethacrylate	142	136
14, 15	4	Toluene	92	93
	5	Styrene	104	109
	11	1,3-Diphenylbutadiene	206	205
	16	2,4,6-Triphenyl-1-hexene	313	339
18, 19	1	1,3-Butadiene	54	56·3
	2	Cyclopentene	68	68·7
	5	1,3-Cyclohexadiene	80	77·2
	9	4-Vinyl-1-cyclohexene	108	108·8

which can abstract a hydrogen atom from another molecule and produce alkanes (**V**).

$$CH_3(CH_2)_{n-2}CH_3$$

(**V**)

Highly strained conformations will not be permitted in the intramolecular backbiting steps. The process is therefore argued to involve preferentially

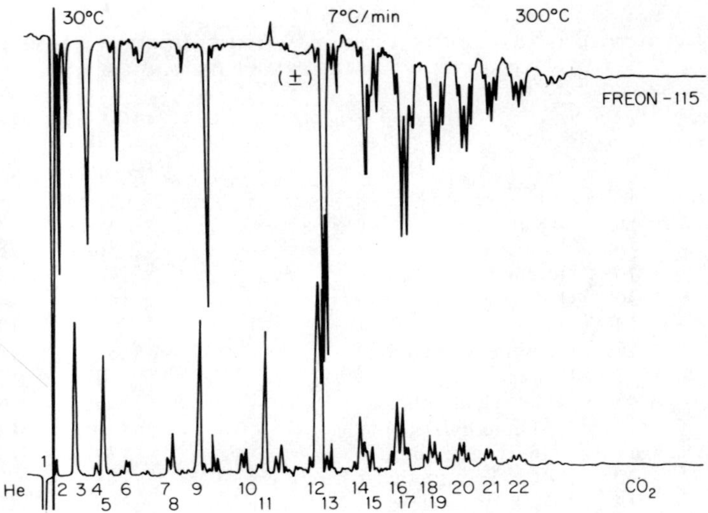

FIG. 7. Mass chromatogram of the volatile pyrolysis products of isotactic polypropylene (~100% isotactic, $\bar{M}_w = 284\,000$ and $\bar{M}_w/\bar{M}_n = 2{\cdot}98$). In the CO_2 channel peak attenuations were ×32 for peak 9, and ×8 for all others. In the Freon-115 channel attenuations were ×128 (peak 1), ×64 (peaks 2 through 5), ×32 (peaks 6 through 11), and ×8 thereafter. Columns (SE-30) were program-heated at 7 °C/min from 30° to 300 °C.

the 5th carbon atom through a pseudo six-membered ring intermediate. If subsequent repetitive intramolecular transfer follows before chain scission, the consequence of preferential radical transfer to the 5th, 9th, 13th, 17th, 21st ... carbon atoms is the preferential formation of C_6, C_{10}, C_{14}, C_{18}, C_{22} ... alkenes (reaction **II → III**) and C_3, C_7, C_{11}, C_{15}, C_{19} ... alkanes (reactions **II → IV → V**), respectively. The probability of sequential migration of the free radical along the chain without undergoing β-scission reactions is expected to decrease as the number of transfer steps increases. One and two step processes should be most dominant and, as observed in Fig. 6, among alkenes, C_6 and C_{10}, and among alkanes, C_3 and C_7 represent major constituents.

The mass chromatographic output obtained for isotactic polypropylene is shown in Fig. 7. Calculation of the molecular weights has shown that the peaks in the chromatogram have molecular weights corresponding to those of the monomer, dimers, trimers and higher oligomers.[1,7] Calculated molecular weights for some of the peaks are included in Table 1. The monomer (peak 1), which is formed extensively, is produced from either the primary or the secondary radicals formed on chain homolysis. Intramolecular radical transfers to the 6th, 10th, and 12th carbon atoms in the primary macroradicals and to the 5th and 9th and also the 13th carbon atoms in the secondary macroradicals (indexing from the secondary carbon radical at the chain end) account for the other major products of decomposition. The formation of the other volatile products of pyrolysis of polypropylene and their relative abundances have been explained also in terms of intramolecular radical transfer processes. In polypropylene initiation by chain scission results in primary (**I**) and secondary (**II**) radicals.

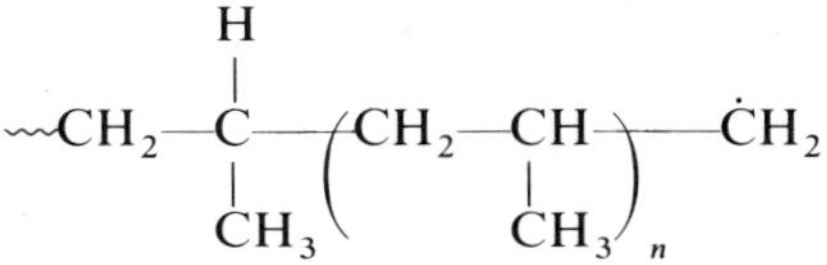

$$\sim CH_2-\underset{CH_3}{\overset{H}{C}}-\left(CH_2-\underset{CH_3}{CH}\right)_n-\dot{C}H_2 \quad \textbf{(I)}$$

$$\sim CH_2-\underset{CH_3}{\overset{H}{C}}-\left(CH_2-\underset{CH_3}{CH}\right)_n-CH_2-\underset{CH_3}{\dot{C}H} \quad \textbf{(II)}$$

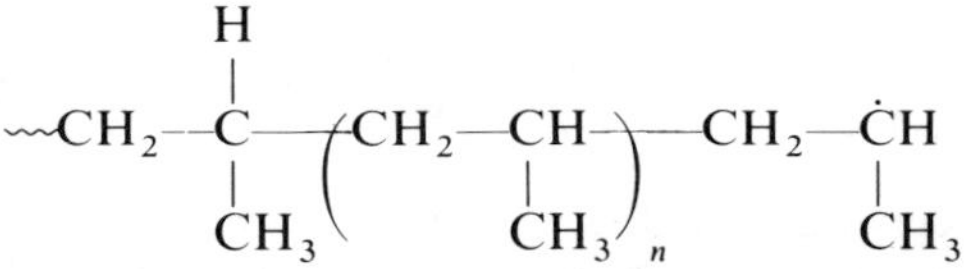

Intramolecular transfer reactions followed by β-scission to the left of the resulting tertiary radical produces alkenes (**III**) from primary radicals and alkenes (**IV**) from secondary radicals.

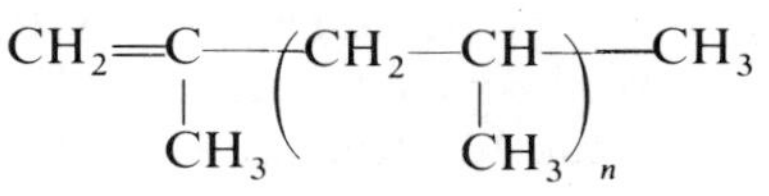

$$CH_2{=}\underset{CH_3}{C}-\left(CH_2-\underset{CH_3}{CH}\right)_n-CH_3 \quad \textbf{(III)}$$

$$CH_2{=}\underset{CH_3}{C}-\left(CH_2-\underset{CH_3}{CH}\right)_n-CH_2-\underset{CH_3}{CH_2} \quad \textbf{(IV)}$$

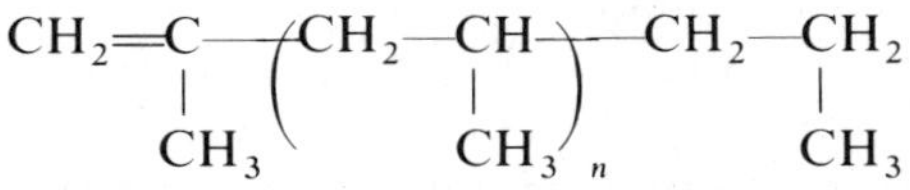

β-Scission to the right of the tertiary radical, followed by hydrogen atom abstraction, leads to the formation of alkanes (**V**) and (**VI**) from primary and secondary macroradicals, respectively.

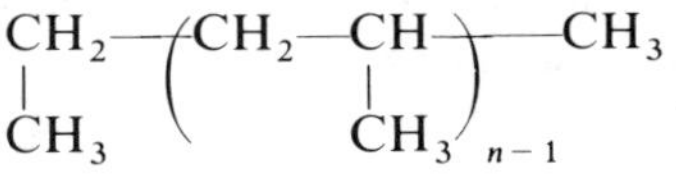

$$\underset{CH_3}{CH_2}-\left(CH_2-\underset{CH_3}{CH}\right)_{n-1}-CH_3 \quad \textbf{(V)}$$

$$\underset{CH_3}{CH_2}-\left(CH_2-\underset{CH_3}{CH}\right)_{n-1}-CH_2-\underset{CH_3}{CH_2} \quad \textbf{(VI)}$$

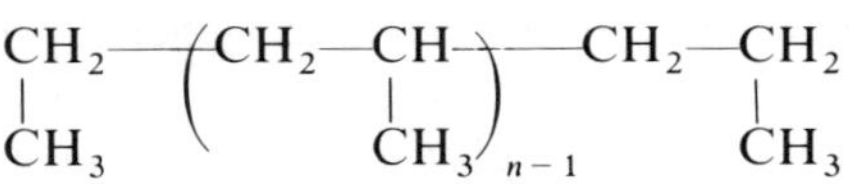

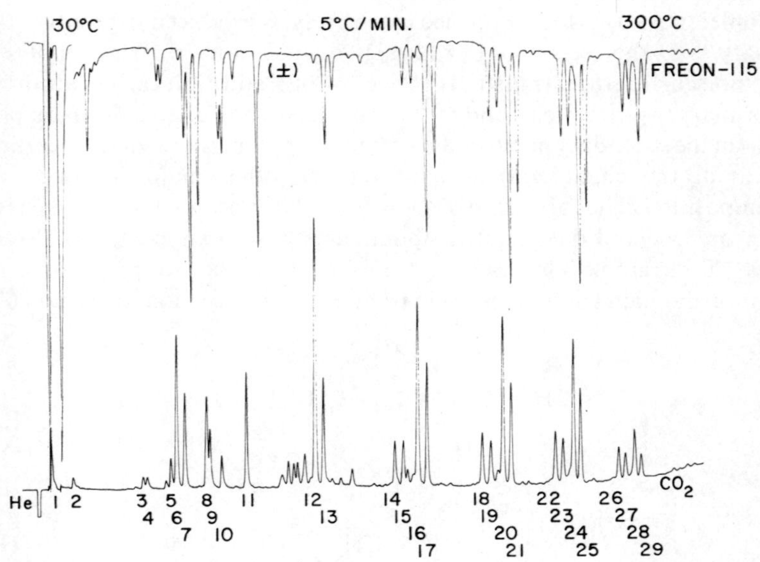

FIG. 8. Mass chromatogram of the volatile pyrolysis products of polyisobutylene ($\bar{M}_v = 120\,000$). Peak attenuations were $\times 8$ in both channels, except for peaks 1 and 2 for which the setting was $\times 128$. Columns (SE-30) were program-heated at 5 °C/min from 30° to 300 °C.

Specific major decomposition products, i.e., pentane (peak 3, structure **VI**, $n = 1$), 2,4-dimethyl-1-heptene (peak 9, structure **IV**, $n = 1$) and 2,4,6,8-tetramethyl-1-hendecene (peak 12, structure **IV**, $n = 3$) can form through intramolecular radical transfer in the secondary radicals, and 2-methylpentane (peak 5, structure **V**, $n = 2$), 2,4,6-trimethylnonane (peak 11, structure **V**, $n = 4$) and 2,4,6,8-tetramethylhendecane (peak 13, structure **V**, $n = 5$) can form from intramolecular radical transfer in the primary radicals. Among the higher molecular weight fragments 2,4,6,8,10,12-hexamethyl-1-tridecene (peak 16) and 2,4,6,8,10,12-hexamethyl-1-pentadecene (peak 17) are formed in larger amounts. Their formation is a continuation of the processes leading to peaks 12 and 13 which are major decomposition products.

The mass chromatogram for polyisobutylene displays a series of regularly spaced groups of peaks which is characteristic of a homologous series of products (Fig. 8). In addition to depolymerisation processes which account for the extensive formation of the monomer (peak 1), intramolecular radical transfer processes in the primary and tertiary macroradicals (which form on initial chain scission) account for the formation of dimers, trimers, and high oligomers.[1,7] More intramolecular transfer occurs in the primary radicals than the tertiary radicals for energetic reasons. A major product, the trimer 2,4,4,6,6-pentamethyl-1-heptene (peak 12), is formed from a primary radical via intramolecular radical transfer to a methyl group:

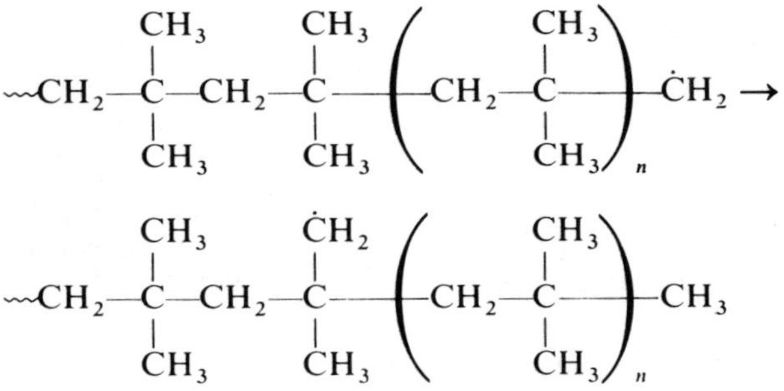

$$\sim CH_2-\underset{CH_3}{\overset{CH_3}{\underset{|}{\overset{|}{C}}}}-CH_2-\underset{CH_3}{\overset{CH_3}{\underset{|}{\overset{|}{C}}}}-\left(CH_2-\underset{CH_3}{\overset{CH_3}{\underset{|}{\overset{|}{C}}}}\right)_n-\dot{C}H_2 \rightarrow$$

$$\sim CH_2-\underset{CH_3}{\overset{CH_3}{\underset{|}{\overset{|}{C}}}}-CH_2-\underset{CH_3}{\overset{\dot{C}H_2}{\underset{|}{\overset{|}{C}}}}-\left(CH_2-\underset{CH_3}{\overset{CH_3}{\underset{|}{\overset{|}{C}}}}\right)_n-CH_3$$

When the radical thus formed undergoes β-scission to the left of the adjacent carbon, alkenes with the structure (**I**) are formed:

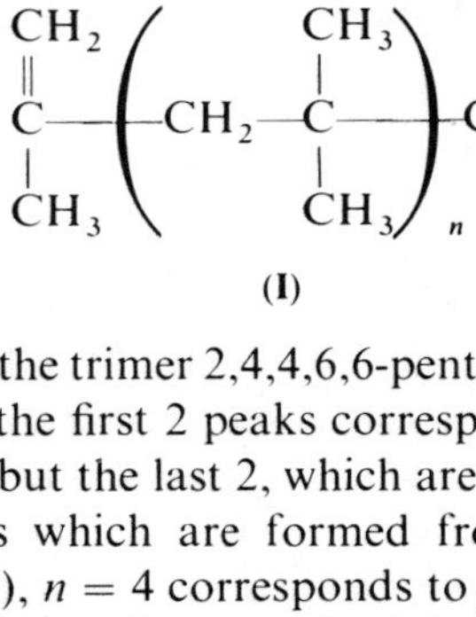

$$\underset{CH_3}{\overset{CH_2}{\underset{|}{\overset{\|}{C}}}}-\left(CH_2-\underset{CH_3}{\overset{CH_3}{\underset{|}{\overset{|}{C}}}}\right)_n-CH_3$$

(**I**)

For $n = 2$ the alkene is the trimer 2,4,4,6,6-pentamethyl-1-heptene. In each quartet (peaks 14–29) the first 2 peaks correspond to alkenes which arise from tertiary radicals, but the last 2, which are formed in larger amounts, correspond to alkenes which are formed from primary radicals. For example in structure (**I**), $n = 4$ corresponds to peak 20, $n = 5$ to peak 24. Peaks 21 and 25 arise from primary radicals but via intramolecular radical transfer to methylene instead of a methyl group. Assigned structures and calculated molecular weights for some of the peaks are given in Table 1.

3.2. Polyolefinsulphones

Poly(butene-1-sulphone), poly(pentene-1-sulphone) and poly(hexene-1-sulphone) have been studied by pyrolysing samples in helium at a heating rate of 20 °C/min.[1,8] The volatile products formed in the temperature range 100–500 °C were analysed by the mass chromatograph.

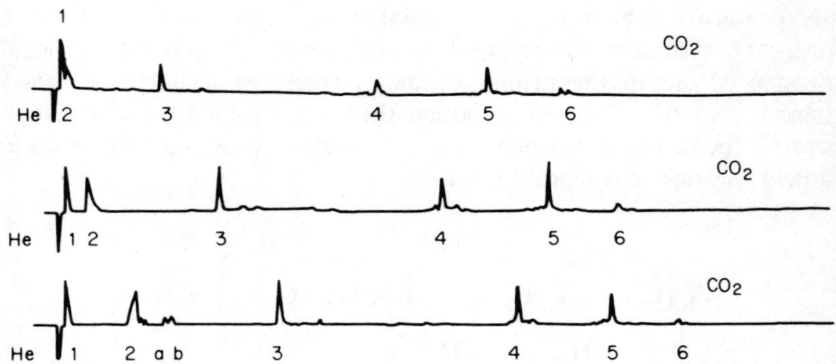

FIG. 9. Comparison of the CO_2 channels of the mass chromatograms of volatile pyrolysis products of poly(butene-1-sulphone) (top), poly(pentene-1-sulphone) (middle), and poly(hexene-1-sulphone) (bottom).

The mass chromatographic outputs are shown for poly(pentene-1-sulphone) in Fig. 3. The results for the three polymers are compared in Fig. 9 in terms of the outputs of the CO_2 channel of the mass chromatograph. The primary products of decomposition have been found to be SO_2 (peak 1) and the respective olefin, i.e., butene, pentene, and hexene (peak 2). These polysulphones contain $-\overset{|}{\underset{|}{C}}-\overset{O}{\overset{\|}{\underset{\underset{O}{\|}}{S}}}-$ linkages in the polymer backbone. The bond dissociation energy for the C—S bond (55–60 kcal/mol) is appreciably lower than for the C—C bond (80–85 kcal/mol) and the C—H bond (90–100 kcal/mol).[13] Consequently thermal decomposition in these polymers may occur readily by scission of the C—S bond and results in the formation of SO_2 and the respective olefin.

As seen in Fig. 9, in going from one chromatogram to the next, the locations of the constituents (peaks 2–6) other than SO_2 (peak 1) shift to higher retention times in an ordered fashion. Calculations of the molecular weights have shown that in each chromatogram the incremental increase in molecular weight in going from peak 2 to 3 is 32 mass units, from peak 3 to 4 is equal to the molecular weight of the respective olefin, from 4 to 5 is 32 mass units, and from 5 to 6 is 16 mass units. Based on the atomic weight of S being 32 and that of O being 16, some compositional possibilities were suggested for peaks 3–6. These have the form $[C_n^{=}][S]$ for peaks 3,

$[C_n{}^{=}]_2[S]$ for peaks 4, $[C_n{}^{=}]_2[S][O]_2$ or $[C_n{}^{=}][S]_2$ for peaks 5, and $[C_n{}^{=}]_2[S][O]_3$ or $[C_n{}^{=}]_2[S]_2[O]$ for peaks 6, where C_n represents C_4 olefin, C_5 olefin or C_6 olefin. Mechanisms for the formation of these structures were not advanced.[1,8] Their formation probably involves some destruction of SO_2 in the presence of free radicals.

These polymers all show a two-step decomposition phenomenon (see Fig. 3). Selective trapping and analysis of the volatile decomposition products which form in the temperature range 100–200 °C showed that only SO_2 and the respective olefin are formed in the first step of degradation.

In addition to these polyolefinsulphones, a sample of polystyrenesulphone has been studied.[1,8] Sulphur dioxide and styrene are found to be the major products of decomposition. Additional fragments derived from styrene dimers were observed as indicated by comparison of the results with those obtained from pyrolysis of polystyrenes; this provided evidence for the presence of consecutive styrene residues in the repeat unit of the polystyrenesulphone.

3.3. Polymethacrylates

Polymethylmethacrylates and poly-t-butylmethacrylates of different tacticities have been studied by pyrolysing samples in a flowing helium atmosphere at a heating rate of 20 °C/min.[1]

Thermal conductivity responses during pyrolysis of polymethylmethacrylates showed some differences for the different tactic forms (Fig. 10). Analysis of the products of pyrolysis formed in the temperature range 200–500 °C shows, however, that all the tactic forms revert almost exclusively to a single compound with a calculated molecular weight which compares well with that of the monomer, methylmethacrylate (Table 1). Mass chromatograms are shown in Fig. 11. Analysis of the products of pyrolysis formed from the isotactic polymer in the temperature range 420–500 °C (see Fig. 10) has shown that in this region monomer formation is no longer observed (see Fig. 11D). The traces of high molecular weight constituents observed were not characterised.

The thermal histories during pyrolysis of the poly-t-butylmethacrylates are shown in Fig. 12. These polymers behave similarly and show two-stage decomposition. The mass chromatogram of the volatile products of pyrolysis formed in the temperature range 140–500 °C is shown in Fig. 13 for the isotactic polymer. The major products of decomposition of all of the tactic forms are found to be isobutylene (peak 1) and the monomer, t-butylmethacrylate (peak 2) (Table 1). Formation of the monomer is a

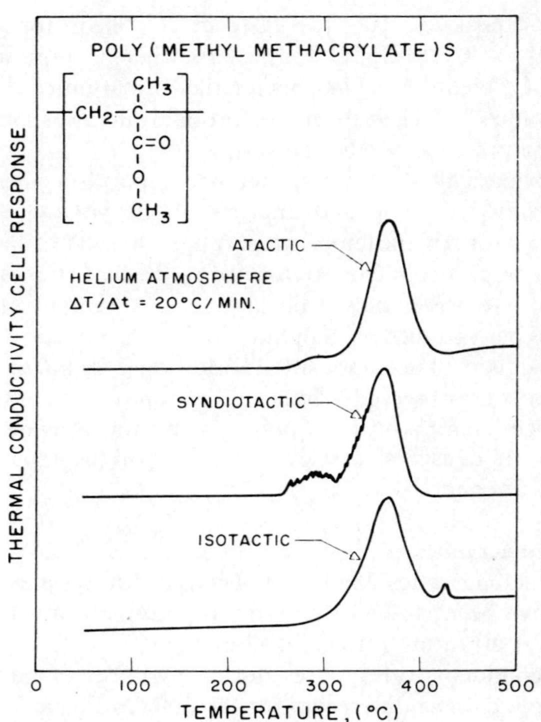

FIG. 10. Thermal conductivity cell responses during pyrolysis of atactic ($\bar{M}_w = 105\,000$, $\bar{M}_w/\bar{M}_n = 2{\cdot}15$), syndiotactic (70% syndio, $\bar{M}_w = 83\,200$, $\bar{M}_w/\bar{M}_n = 1{\cdot}33$) and isotactic (91·5% iso, $\bar{M}_w = 2\,780\,500$, $\bar{M}_w/\bar{M}_n = 2{\cdot}28$) polymethylmethacrylates.

consequence of depolymerisation. The formation of isobutylene is indicative of ester decomposition, i.e.,

$$\sim\!\!\underset{\substack{|\\ C=O\\ |\\ O\\ |\\ C(CH_3)_3}}{\overset{\substack{CH_3\\ |}}{C}}\!-\!CH_2\!-\!\underset{\substack{|\\ C=O\\ |\\ O\\ |\\ C(CH_3)_3}}{\overset{\substack{CH_3\\ |}}{C}}\!-\!CH_2\!\sim \;\longrightarrow\; \sim\!\!\underset{\substack{|\\ C\\ \diagup\!\!\diagup\quad\diagdown\\ O\qquad OH}}{\overset{\substack{CH_3\\ |}}{C}}\!-\!CH_2\!-\!\underset{\substack{|\\ C\\ \diagup\quad\diagdown\!\!\diagdown\\ OH\qquad O}}{\overset{\substack{CH_3\\ |}}{C}}\!-\!CH_2\!\sim$$

$$+\; 2H_2C{=}C(CH_3)_2$$

Isobutylene

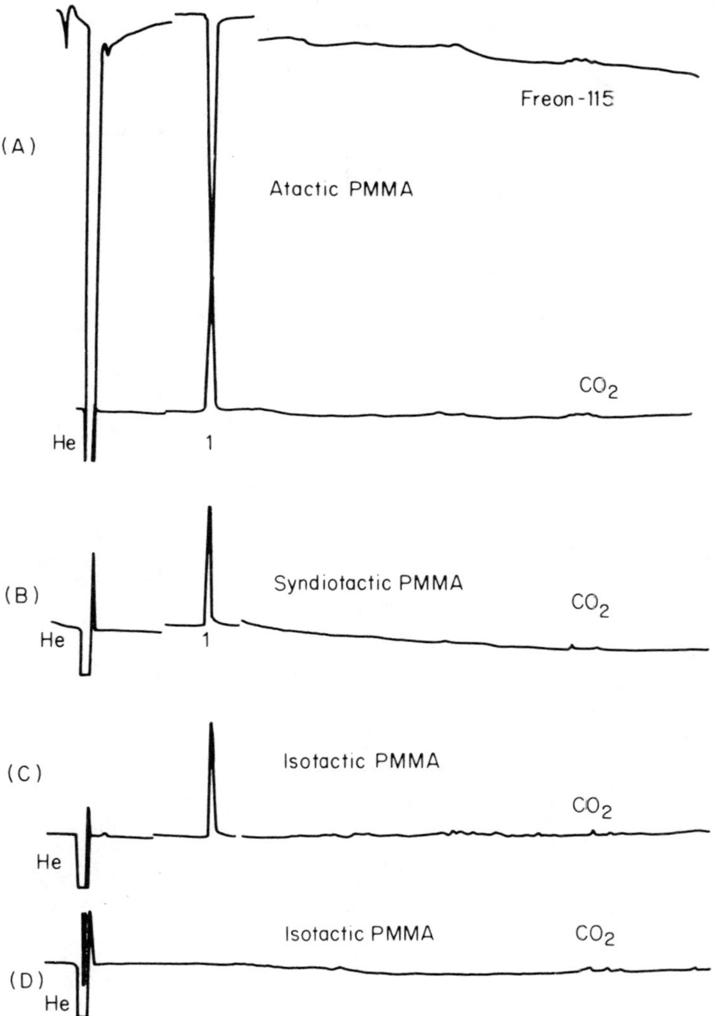

FIG. 11. Mass chromatogram of the volatile products of pyrolysis formed in the temperature range 200–500 °C of atactic (A), syndiotactic (B) and isotactic (C) polymethylmethacrylates. Only the CO_2 channel outputs are shown for the syndiotactic and isotactic polymers. Columns (Dexsil-300) were program-heated at 5 °C/min from 30 to 300 °C. Peak attenuations were × 64 for peak 1, and × 8 for all others. The figure at the bottom (D) shows the CO_2 channel output for the products formed from the isotactic polymer in the temperature range 420–500 °C.

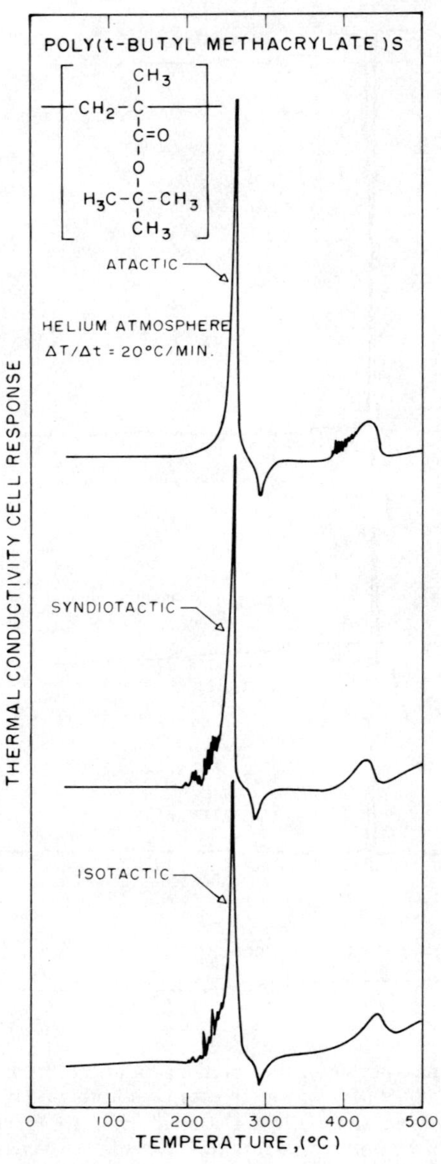

FIG. 12. Thermal conductivity cell responses during pyrolysis of atactic ($\bar{M}_w = 359\,950$, $\bar{M}_w/\bar{M}_n = 2{\cdot}03$), syndiotactic ($\bar{M}_w = 20\,940$, $\bar{M}_w/\bar{M}_n = 1{\cdot}49$) and isotactic (82% iso, $\bar{M}_w = 300\,000$, $\bar{M}_w/\bar{M}_n = 1{\cdot}13$) poly-t-butylmethacrylates.

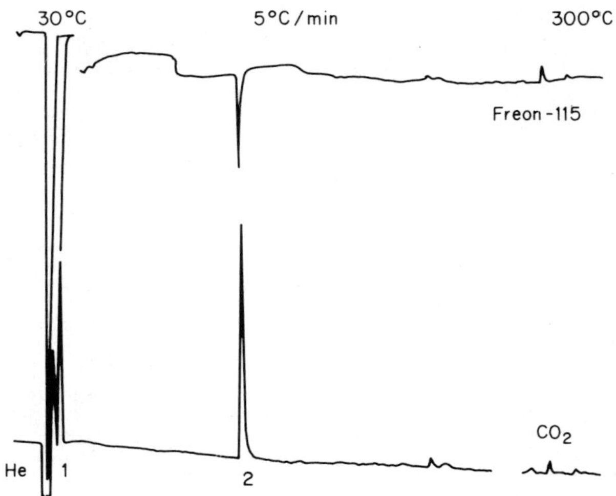

FIG. 13. Mass chromatogram of the volatile products of pyrolysis of isotactic poly-t-butylmethacrylate. Columns (Dexsil-300) were program-heated at 5 °C/min from 30° to 300 °C. Peak attenuations were × 8, except for peak 1 for which the settings were × 128 in the Freon channel and × 16 in the CO_2 channel.

The relative extents of these reactions were found to be dependent on the tacticity of the sample as evidenced by differences in the monomer to isobutylene peak height ratios measured in the outputs of the CO_2 channel of the mass chromatograph.[1] Monomer to isobutylene ratio was found to be smallest for the syndiotactic and largest for the isotactic polymer. This observation was interpreted to mean that the extent of depolymerisation compared with ester decomposition is greatest with the isotactic polymer; the steric hindrances of the bulky side-groups apparently favour depolymerisation most in the isotactic polymer. Further analyses of the volatile decomposition products formed in the temperature ranges 140–225 °C, 140–320 °C and 320–500 °C in the isotactic polymer showed that all monomer generation takes place in the temperature range 140–320 °C. In the second stage of decomposition (i.e., from 320 to 500 °C) no monomer is generated, but some isobutylene continues to form.

3.4. Polystyrenes

Anionic, thermal, and comb-shaped polystyrenes have been studied.[3,9]

Anionic polystyrenes with different average molecular weights, which had been polymerised using n-butyl lithium as the initiator and terminated

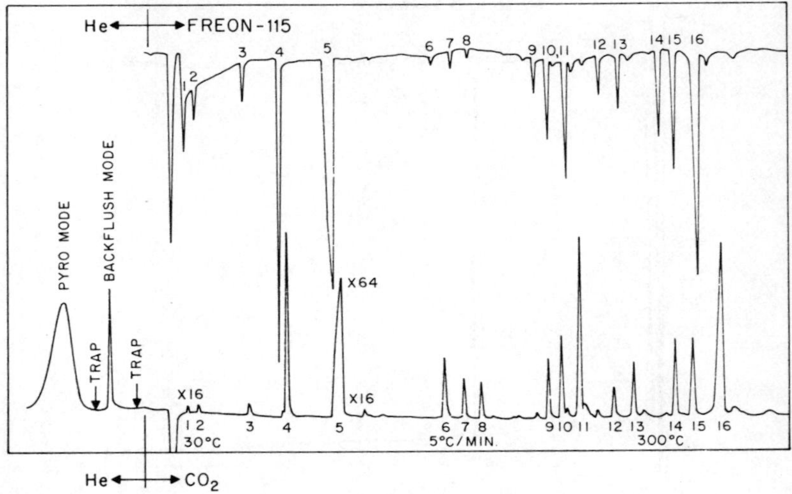

FIG. 14. Mass chromatogram of the volatile products of pyrolysis of anionic polystyrene ($\bar{M}_n = 2100$, $\bar{M}_w/\bar{M}_n < 1{\cdot}1$) with n-butyl ends. Columns (Dexsil-300) were program-heated at 5 °C/min from 30° to 300 °C.

with protons, were pyrolysed in flowing helium at a heating rate of 20 °C/min, and the products volatilising in the temperature range 200–500 °C were analysed. The mass chromatographic output for a low molecular weight polymer is shown in Fig. 14. Calculations of the molecular weights and examination of the on-the-fly IR spectra showed that the major decomposition products are toluene (peak 4), styrene (peak 5), 1,3-diphenylbutadiene (peak 11) and 2,4,6-triphenyl-1-hexene (peak 16) (Tables 1 and 2). The relative amounts of these constituents have been found to depend on the molecular weight of the initial polymer. Among the constituents, toluene shows the highest sensitivity to molecular weight (Table 2).

In polystyrene, cleavage of the backbone results in the primary (**I**) and secondary (**II**) radicals.

$$\sim\underset{\phi}{\overset{\text{H}}{\text{C}}}-\dot{\text{C}}\text{H}_2 \qquad\qquad \cdot\underset{\phi}{\overset{\text{H}}{\text{C}}}-\text{CH}_2-\left(\underset{\phi}{\text{CH}}-\text{CH}_2\right)_n-\underset{\phi}{\text{CH}}\sim$$

(I) **(II)**

TABLE 2

VOLATILE PRODUCTS OF PYROLYSIS OF ANIONIC AND THERMAL POLYSTYRENES. VARIATION OF THEIR RELATIVE AMOUNTS (AS WEIGHT PERCENTAGES) WITH THE MOLECULAR WEIGHT OF THE INITIAL POLYMER

Peak number/$\bar{M}_n$	*Anionic*					*Thermal*		
	2 100	4 000	10 000	20 400	390 000	1 880	6 270	111 000
1	0·076	0·031	Trace	Trace	0	0	0	0
2	0·69	0·37	Trace	Trace	0	0	0	0
3	0·71	0·38	1·77	2·97	1·82	0	0·32	0
4	8·33	5·24	2·79	1·99	1·31	1·0	1·40	2·54
5	37·4	35·8	41·2	45·7	49·0	42·80	48·00	50·90
6	4·14	1·61	Trace	Trace	0	0	0	0
7	2·72	3·64	Trace	Trace	0	0	0	0
8	2·56	1·14	Trace	Trace	0	0	0	0
9	3·02	2·75	3·39	3·04	2·22	0	0	0
10	4·11	5·25	5·75	4·36	5·91	Trace	Trace	Trace
11	8·96	9·76	12·56	11·74	11·37	17·50	18·00	18·87
12	1·66	1·17	0·50	Trace	0	0	0	0
13	2·99	2·23	1·27	Trace	0	0	0	0
14	4·53	5·85	6·45	7·19	5·07	0	0	0
15	5·75	7·63	7·38	6·57	8·69	0	0	0
16	22·33	18·14	16·70	16·44	14·64	35·60	29·80	26·33

Both of these can undergo depolymerisation and form styrene (peak 5, Tables 1 and 2).

1,3-Diphenylbutadiene (peak 11, Tables 1 and 2) forms from tertiary radicals of the form:

$$\sim CH_2-\underset{\phi}{\underset{|}{\dot{C}}}-CH_2\sim$$

formed by hydrogen abstraction, which cleave β from the tertiary radical to form terminally unsaturated species with allylic hydrogens:

$$\sim CH_2-\underset{\phi}{\underset{|}{\dot{C}}}-CH_2-\underset{\phi}{\underset{|}{CH}}\sim \rightarrow \sim CH_2-\underset{\phi}{\underset{|}{C}}=CH_2+\cdot\underset{\phi}{\underset{|}{CH}}\sim$$

Abstraction of the allylic hydrogen followed by β-scission of the resulting tertiary radical leads to 1,3-diphenylbutadiene:

$$\sim CH_2-\underset{\phi}{\underset{|}{CH}}-\dot{C}H-\underset{\phi}{\underset{|}{C}}=CH_2 \rightarrow \sim CH_2\cdot+\underset{\phi}{\underset{|}{CH}}=CH-\underset{\phi}{\underset{|}{C}}=CH_2$$

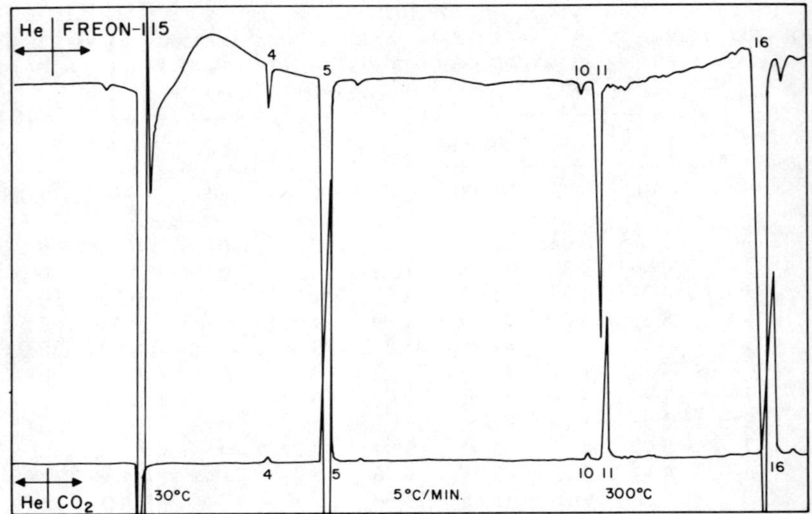

FIG. 15. Mass chromatogram of the volatile products of pyrolysis of thermal polystyrene ($\bar{M}_n = 1880$, $\bar{M}_w/\bar{M}_n = 1{\cdot}01$). Columns (Dexsil-300) were program-heated at 5 °C/min from 30° to 300 °C.

Intramolecular radical transfer in secondary radicals of type **II**, via a pseudo six-membered ring intermediate, may result in tertiary radicals such as:

$$\sim\underset{\phi}{\underset{|}{\mathrm{CH}}}-\mathrm{CH_2}-\underset{\phi}{\underset{|}{\dot{\mathrm{C}}}}-\mathrm{CH_2}-\underset{\phi}{\underset{|}{\mathrm{CH}}}-\mathrm{CH_2}-\underset{\phi}{\underset{|}{\mathrm{CH_2}}}$$

which upon β-scission to the left of the radical centre leads to 2,4,6-triphenyl-1-hexene (peak 16, Tables 1 and 2):

$$\mathrm{CH_2}{=}\underset{\phi}{\underset{|}{\mathrm{C}}}-\mathrm{CH_2}-\underset{\phi}{\underset{|}{\mathrm{CH}}}-\mathrm{CH_2}-\underset{\phi}{\underset{|}{\mathrm{CH_2}}}$$

Depolymerisation reactions in secondary radical **II**, if not stopped by other mechanisms, may eventually lead to the primary radical $\phi-\dot{\mathrm{C}}\mathrm{H_2}$ which can abstract a hydrogen atom from another molecule to yield toluene (peak 4, Tables 1 and 2).

Thermally polymerised polystyrenes have been found to display a different decomposition pattern from that of anionic polystyrenes. Comparison of the mass chromatograms shown in Figs. 14 and 15 and examination of Table 2 shows that fewer products are formed with thermal polystyrenes; those peaks which are highly molecular weight dependent (i.e., influenced by chain ends) are absent.

Another set of low molecular weight anionic polystyrenes initiated with a dicarbanionic species and terminated at both ends with either proton, α-naphthyl, diphenylmethyl or hydroxyl groups, were also studied. These

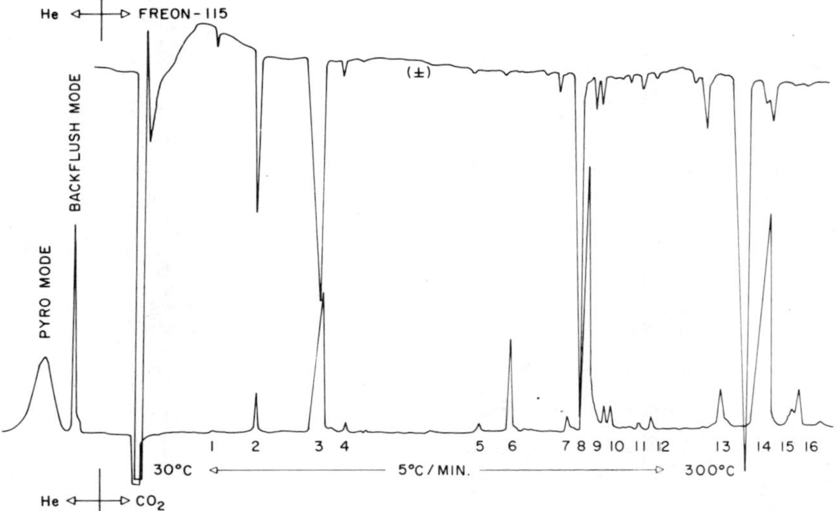

FIG. 16. Mass chromatogram of the volatile products of pyrolysis of anionic polystyrene ($\bar{M}_n = 10\,500$, $\bar{M}_w/\bar{M}_n = 1{\cdot}15$) with α-naphthyl ends. Columns (Dexsil-300) were program-heated at 5 °C/min from 30° to 300 °C.

polymers have been found to give products of decomposition which are similar to those of thermal polystyrenes. Polymers with α-naphthyl ends and diphenylmethyl ends showed some differences in that products arising from these groups are also observed. Figure 16 shows the mass chromatogram for the polymer with α-naphthyl ends. Peaks 5 and 6, predicted to be naphthylethane and 3-naphthyl-1-propene, respectively, are characteristic of this polymer.

The major products of pyrolysis of comb-shaped polystyrenes were found to be similar to those of thermal polystyrenes also. However, the steric hindrance of the side branches in the comb-shaped polymers was considered to affect the relative abundance of the observed constituents. Formation of the trimer (peak 16) is restricted and lesser amounts form.

3.5. Polybutadienes

1,4-Polybutadienes with different *cis*/*trans* ratios have been studied.[10]

Thermal histories before and during pyrolysis of a high *trans* and a high *cis*-1,4-polybutadiene in an inert atmosphere are shown in Fig. 17. The polymers undergo two-step decomposition. Degradation to 15% weight loss and also after 15% weight loss has been investigated. Mass

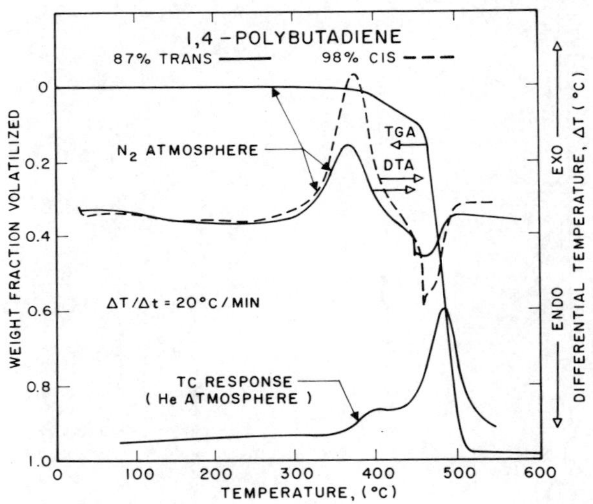

FIG. 17. Thermal history before and during pyrolysis of 1,4-polybutadienes. Solid lines for sample with 87% *trans*, $\bar{M}_w = 67\,300$, $\bar{M}_w/\bar{M}_n = 2{\cdot}18$; and dashed lines for sample with 0% *trans*, $\bar{M}_w = 633\,500$, $\bar{M}_w/\bar{M}_n = 3{\cdot}62$.

chromatograms of the volatile products formed during 15% weight degradation from a high *cis* and a high *trans* sample are shown in Figs. 18 and 19, respectively. Calculation of the molecular weights and examination of on-the-fly IR spectra (Fig. 20) have shown that the major products of decomposition are 1,3-butadiene (monomer, peak 1), cyclopentene (peak 2), 1,3-cyclohexadiene (peak 5) and 4-vinyl-1-cyclohexene (dimer, peak 9). Their relative amounts were found to vary with the *cis*/*trans* ratio in the polymer sample. The relative quantities of 4-vinyl-1-cyclohexene and 1,3-butadiene decrease, cyclopentene increases and 1,3-cyclohexadiene is relatively unchanged with increasing *trans* content as shown in Fig. 21.

The mass chromatogram of the products of pyrolysis after 15% weight

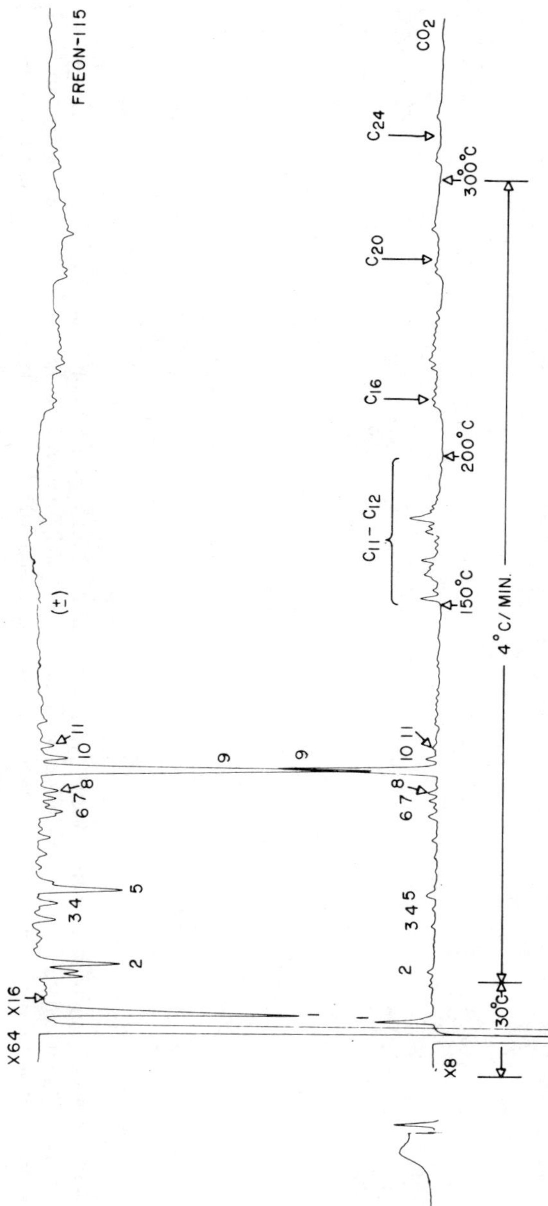

FIG. 18. Mass chromatogram of the volatile pyrolysis products of high *cis*-1,4-polybutadiene (0% *trans*, $\bar{M}_w = 633\,500$, $\bar{M}_w/\bar{M}_n = 3{\cdot}62$) formed during 15% weight loss. Columns (Dexsil-300) were program-heated at 4 °C/min from 30° to 300 °C. Peak attenuations for the Freon-115 channel were $\times 64$ for peak 1, and $\times 16$ thereafter; that for the CO_2 channel was $\times 8$ for all peaks.

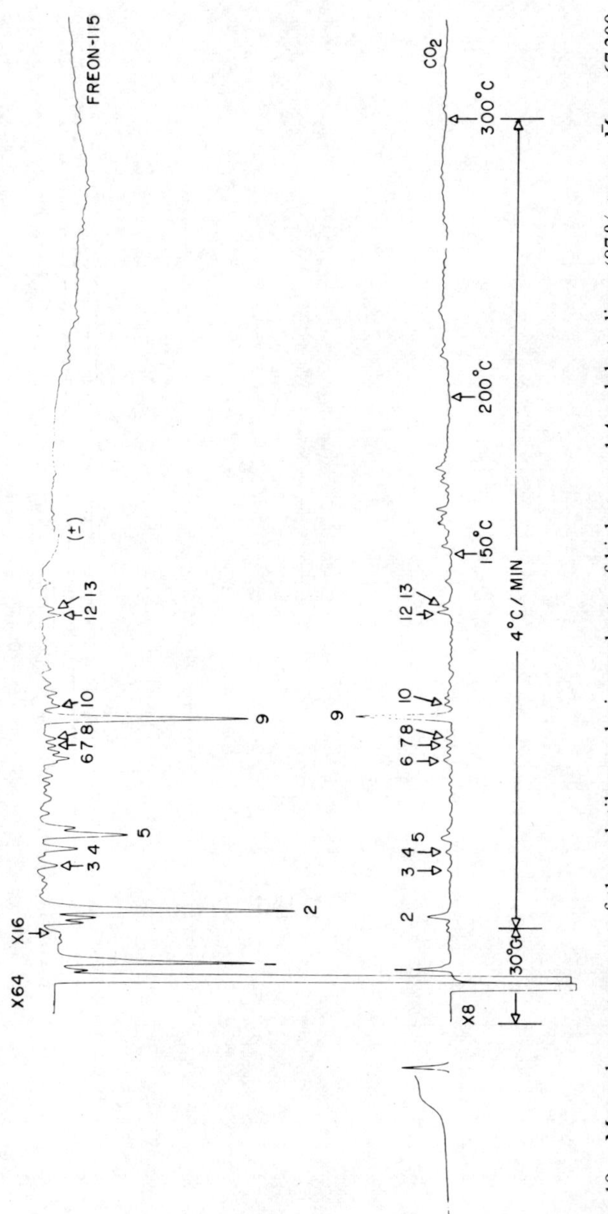

FIG. 19. Mass chromatogram of the volatile pyrolysis products of high *trans*-1,4-polybutadiene (87% *trans*, $\bar{M}_w = 67\,300$, $\bar{M}_w/\bar{M}_n = 2{\cdot}18$) formed during 15% weight loss. Columns (Dexsil-300) were program-heated at 4 °C/min from 30° to 300 °C. Peak attenuations for the Freon-115 channel were × 64 for peak 1, and × 16 thereafter; that for the CO_2 channel was × 8 for all peaks.

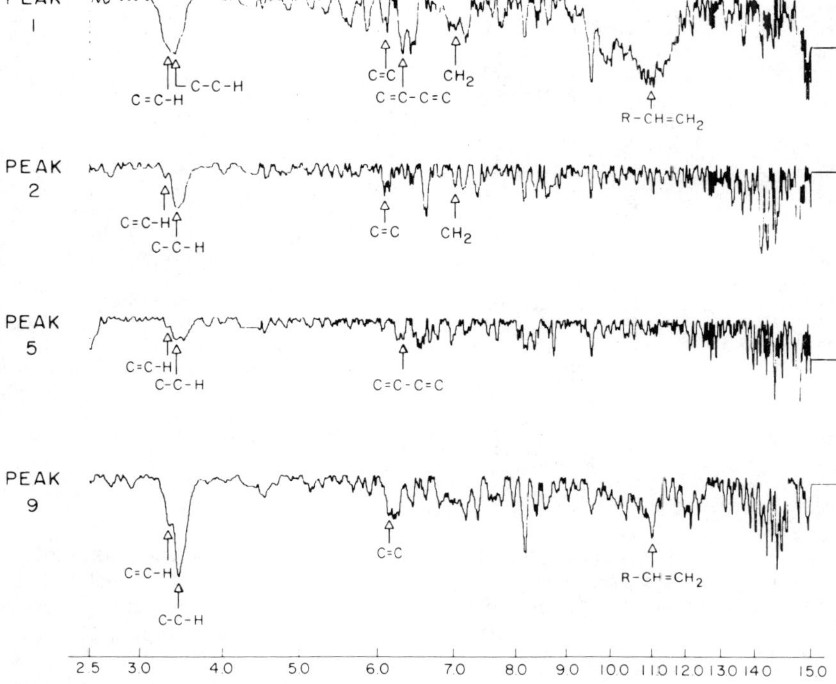

FIG. 20. On-the-fly infrared spectra of peaks 1, 2, 5 and 9 of the pyrogram shown in Fig. 19.

loss of high *trans*-1,4-polybutadiene (Fig. 22) showed that 4-vinyl-1-cyclohexene, which is the most abundant product in degradation to 15% weight loss (see Fig. 19), is only a minor product of further decomposition.

The major products of decomposition of polybutadienes are accounted for by a free radical mechanism. Homolytic scission of the single bond of the polymer chain at a site which is β to the double bond results in the radical:

$$\sim\!CH_2{-}CH{=}CH{-}\dot{C}H_2 \leftrightarrow \sim\!CH_2{-}\dot{C}H{-}CH_2{=}CH_2$$

(I)

The monomer 1,3-butadiene arises from **I** via β-scission, i.e.,

$$\mathbf{I} \rightarrow \sim\!\dot{C}H_2 + CH_2{=}CH{-}CH{=}CH_2$$

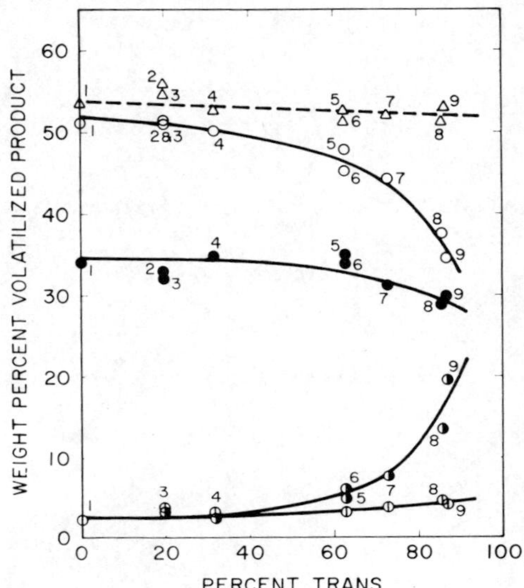

FIG. 21. Volatile degradation products from 1,4-polybutadienes: amounts of main products versus *trans* content (15% weight loss). ●, 1,3-Butadiene; ◑, cyclopentene; ⦶, 1,3-cyclohexadiene; ○, 4-vinyl-1-cyclohexene; △, cyclopentene + 4-vinyl-1-cyclohexene. (Numbers identify samples.[10])

4-Vinyl-1-cyclohexene arises via a backbiting step involving a pseudo six-membered ring intermediate which is favoured by a *cis*-1,4-structure, as is indicated by:

$$\begin{array}{l} \quad\quad\quad H \quad\quad H \\ \quad\quad\quad\quad C{=}C \\ \sim CH_2{-}CH_2 \quad\quad CH_2 \\ \quad\quad\quad \cdot CH{-}CH_2 \\ \quad\quad\quad\quad | \\ \quad\quad\quad\quad CH \\ \quad\quad\quad\quad \| \\ \quad\quad\quad\quad CH_2 \end{array} \rightarrow \sim\dot{C}H_2 + \text{(4-vinylcyclohexene ring with } CH{=}CH_2\text{)}$$

1,2-Vinyl-polybutadiene has also been studied.[10] In contrast to 1,4-polybutadienes, this polymer undergoes one-step decomposition. The mass chromatogram of all of the volatile products of pyrolysis was similar to that

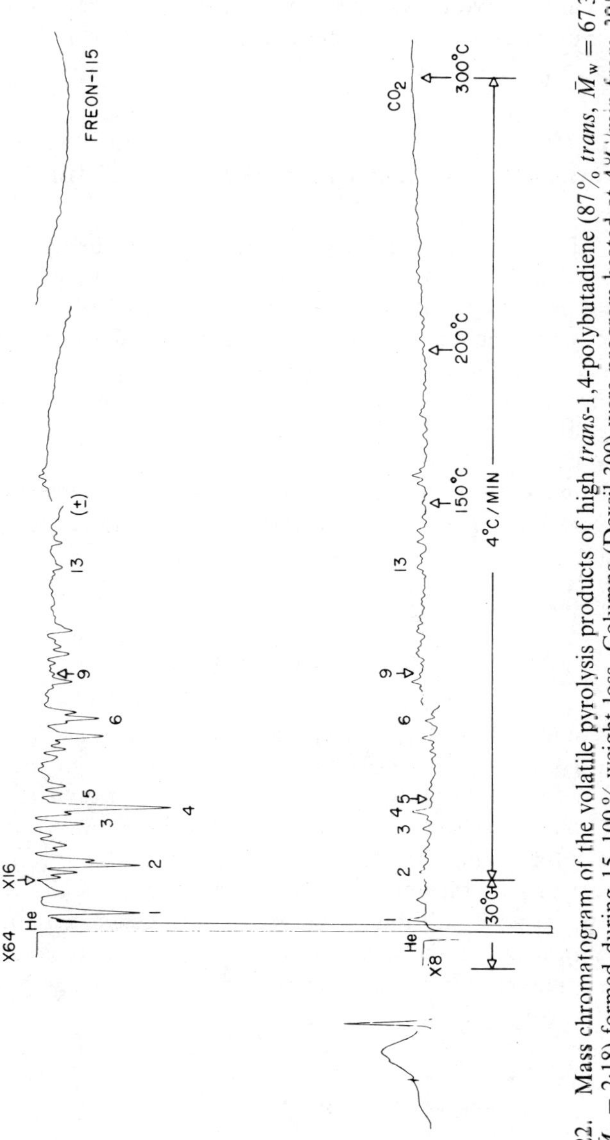

FIG. 22. Mass chromatogram of the volatile pyrolysis products of high *trans*-1,4-polybutadiene (87% *trans*, $\bar{M}_w = 67\,300$, $\bar{M}_w/\bar{M}_n = 2{\cdot}18$) formed during 15–100% weight loss. Columns (Dexsil-300) were program-heated at 4 °C/min from 30° to 300 °C. Peak attenuations for the Freon-115 channel were × 64 for peak 1, and × 16 thereafter; that for the CO_2 channel was × 8 for all peaks.

from pyrolysis above 15% weight loss for 1,4-polybutadienes (compare Figs. 4 and 22). This observation has been interpreted to suggest that part of the 1,4-structure undergoes isomerisation to the 1,2-structure in the course of being heated to 15% weight loss.

4. COMMENTS ON CALCULATED MASS NUMBERS

The accuracy of the molecular weight determination using a mass chromatograph depends on the invariance of the instrument constant and the accuracy and precision with which the response ratios are measured.[11]

Table 1 shows the theoretical molecular weights and the molecular weights calculated from the mass chromatographic outputs for some of the assigned structures in the pyrograms discussed in the foregoing sections. As seen from this table, the calculated values are acceptable values, and thus the mass chromatograph offers a convenience not found in the usual gas chromatograph–mass spectrometer combinations. Mass spectrometry, even though more accurate, requires elaborate interfacial systems and sophisticated data-reduction processes.

5. CONCLUDING REMARKS

The results presented in the foregoing sections show the utility and power of the on-line pyrolysis system in polymer analysis. The system provides thermal history before and during pyrolysis and makes selective trapping and analysis of volatile constituents possible. Chromatographic retention times, easily calculable mass numbers, computer-added IR spectra—all available from an on-line system—facilitate identification of the volatile constituents. Thus not only are fingerprint pyrograms generated, but also detailed information about the products of pyrolysis and their relative abundances are obtained. Based on these, mechanisms for decomposition become easier to propose and to verify.

ACKNOWLEDGEMENT

Partial support from the Office of Naval Research is acknowledged.

REFERENCES

1. Kiran, E., Ph.D. Thesis, Department of Chemical Engineering, Princeton University, Princeton, New Jersey, 1974.
2. Kiran, E. and Gillham, J. K., *J. Appl. Polym. Sci.*, **20** (1976), p. 931.
3. Kuo, H. H., Ph.D. Thesis, Department of Chemical Engineering, Princeton University, Princeton, New Jersey, 1976.
4. Kiran, E. and Gillham, J. K., *Am. Chem. Soc. Polymer Preprints*, **14** (1973), p. 580.
5. Kiran, E. and Gillham, J. K., *Soc. Plastics Engineers, Tech. Papers*, **19** (1973), p. 502.
6. Kiran, E. and Gillham, J. K., *J. Macromol. Sci.-Chem.*, **A8** (1974), p. 211.
7. Kiran, E. and Gillham, J. K., *J. Appl. Poly. Sci.*, **20** (1976), p. 2045.
8. Kiran, E., Gillham, J. K. and Gipstein, E., *J. Appl. Poly. Sci.*, **21** (1977), p. 1159.
9. Kuo, H. H., Pfeffer, H. A. and Gillham, J. K., *Am. Chem. Soc. Coatings and Plastics Preprints*, **35** (1975), p. 434.
10. Tamura, S. and Gillham, J. K., *J. Appl. Poly. Sci.*, **22** (1978), p. 1867.
11. Kiran, E. and Gillham, J. K., *Anal. Chem.*, **47** (1975), p. 983.
12. Stivala, S. S., and Gabbay, S. M., In *Aspects of Degradation and Stabilization of Polymers*, H. H. G. Jellinek, Ed., New York, Elsevier, 1978, Chapter 13.
13. Cottrell, T. L., *The Strength of Chemical Bonds*, London, Butterworth, 1958.

Chapter 2

APPLICATION OF THE LIGHT SCATTERING DETECTION METHOD TO PROBLEMS OF POLYMER DEGRADATION

WOLFRAM SCHNABEL

Hahn-Meitner-Institut für Kernforschung Berlin GmbH, Berlin, West Germany

SUMMARY

The light scattering detection method has been used recently at the Hahn-Meitner-Institut in combination with flash photolysis and pulse radiolysis of polymer solutions in order to elucidate kinetics and mechanisms of polymer degradation. As a consequence of main-chain scission the light scattering intensity (LSI) of polymer solutions is reduced. The time dependence of the decrease of LSI yields in principal two kinds of information which are covered in this chapter by the headings 'Dynamic Studies' and 'Studies of Chemical Reactions'. Dynamic studies include such topics as disentanglement diffusion and preferential solvation, whereas studies of chemical reactions concern the lifetimes and reactions of macroradicals.

1. INTRODUCTION

During recent years research concerning rapid reactions of polymers has been devoted to specific macromolecular topics such as intramolecular reactions and interactions, conformational influences on reaction rates, interpenetration and disentanglement of coiled macromolecules. A wealth of valuable information could be obtained in these directions from absorption and emission measurements.

Since many chemical reactions of macromolecules cause changes in molecular size it appeared to be interesting to measure these changes in order to establish correlations with absorption and emission data and to obtain additional information not otherwise available. Rayleigh light scattering has proved to be an appropriate tool for this purpose.[1]

In this chapter results shall be reviewed which were obtained during the last five years in the author's laboratory. The work essentially concerned processes involving main-chain scission of macromolecules brought about by irradiation of dilute polymer solutions with impulses of high energy electrons or flashes of u.v. light. The duration of the irradiation varied between about 20 ns and several μs. Changes in the light scattering intensity after irradiation were monitored as a function of time. As will be demonstrated below, the light scattering detection method proved to be rather versatile with respect to the various kinds of problems which could be studied.

2. GENERAL ASPECTS

2.1. Mechanistic Considerations

If main-chain scission is the only radiation-induced process leading to a change in the size of the macromolecule (in this case a diminution), the following has to be considered for dilute solutions, where the macromolecules are isolated from each other. Upon main-chain scission the number of macromolecules in the irradiated system is increased while their size is diminished. Thus, immediately after irradiation of a polymer solution with a very short electron pulse or u.v. flash, the spatial distribution of the majority of (freshly generated) macromolecules refers to a non-equilibrium state. As a consequence an entropic force exists which induces the separation of the fragments (as depicted in Fig. 1) and causes the macromolecules to attain a new spatial equilibrium distribution. The diminution of the size of the macromolecules leads to a decrease of the light scattering intensity (LSI). The variation of the LSI can be monitored with a device of appropriate time resolution.

A typical oscilloscope trace demonstrating the LSI decrease after the irradiation of polystyrene in chloroform solution is shown in Fig. 2. In the interpretation of such curves it is important to know that the process of main-chain rupture in solution must be considered as a series of consecutive reactions, where the slowest step is rate determining. This is demonstrated in Fig. 3 where for the sake of simplicity only essential steps of this series are represented schematically. In a primary step (a) the absorption of energy leads to the formation of an activated repeating unit which gives rise to the

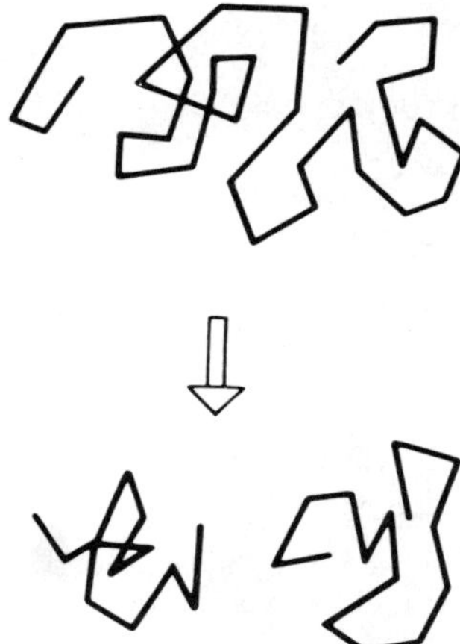

FIG. 1. Schematic representation of the diffusion of fragments after main-chain rupture.

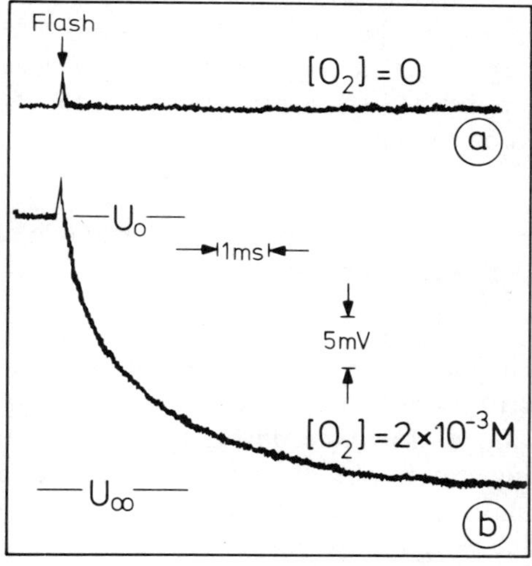

FIG. 2. Degradation of polystyrene in chloroform ($9{\cdot}3 \times 10^{-4}$ base mol/litre) at room temperature; $\bar{M}_{w,0} = 2 \times 10^6$. Oscilloscope traces demonstrating the change of the LSI at $\vartheta = 90°$ after irradiation with a 20 ns flash of 265 nm light. (a) No oxygen present in the solution. (b) $[O_2] = 2 \times 10^{-3}$ mol/litre. Dose absorbed by the solution per flash $= 3{\cdot}2 \times 10^{-6}$ Einstein/litre.[22]

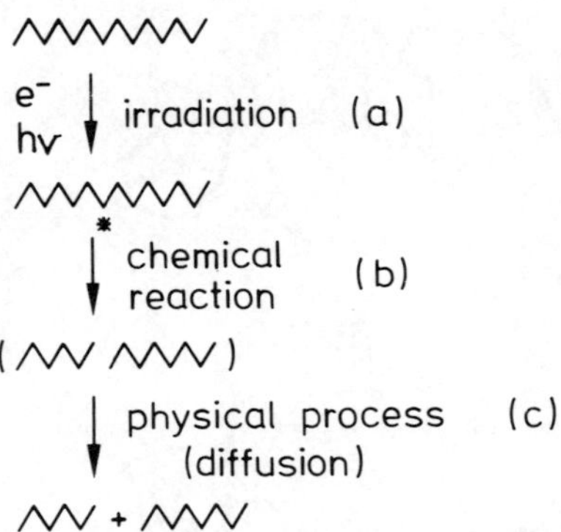

FIG. 3. Scheme demonstrating in a simplified mode the series of consecutive processes during main-chain scission of a macromolecule.

rupture of a bond in the main chain in a succeeding step (b), which is denoted as the 'chemical step'. Bond rupture is followed by step (c) involving the separation of the fragments. This step is denoted as the 'physical step' since it is related to a purely physical process, the diffusion of the scission fragments. In the case of linear flexible macromolecules of coiled conformation, step (c) pertains to disentanglement diffusion.

The activated state generated in step (a) might be either an excited state, a free radical or an ion. Excited states can be formed, for example, by the absorption of a photon by the polymer (P) according to reaction (1):

$$P + h\nu \rightarrow P^* \tag{1}$$

Free macroradicals (P·) can be formed, for example, by hydrogen abstraction according to reaction (2):

$$X\cdot + PH \rightarrow XH + P\cdot \tag{2}$$

Reactive species X· might be formed during the radiolysis or photolysis of the solvent or of a sensitiser which has been added to the system. Charged species are, for example, generated during the primary and secondary ionisations occurring during the absorption of high energy radiation according to reactions (3) and (4):

$$P \rightsquigarrow P^{\cdot +} + e^-_{kin} \tag{3}$$

$$e^-_{kin} + P \rightarrow P^{\cdot +} + 2e^-_{kin} \tag{4}$$

or photochemically, e.g. via the dissociation of exciplexes:

$$(P\text{---}A)^* \rightarrow P^{\cdot +} + A^{\cdot -} \tag{5}$$

From Fig. 3 it can be seen that two extreme cases have to be distinguished: (i) The rate of step (b) is much greater than the rate of step (c) ($v_b \gg v_c$). In this case the lifetime of intermediates formed during, or immediately after, the electron impulse or the u.v. flash is much shorter than the time for the separation of fragments of the macromolecules. The observed rate of LSI decrease is then due to the rate of separation of fragments (in the case of coiled macromolecules due to disentanglement diffusion). (ii) The rate of step (b) is much smaller than the rate of step (c) ($v_b \ll v_c$). In this case the lifetime of the intermediate causing main-chain breakage is much longer than the time of fragment separation. The observed LSI decrease is correlated to the lifetime of the intermediate, i.e. to a chemical reaction. It can, therefore, be concluded that time-resolved light-scattering measurements in combination with rapid means of inducing chemical changes can be used either to study physical phenomena, such as the disentanglement of macromolecules, or to study the kinetics of chemical reactions of macromolecules. As an example of the latter, investigations might be referred to those concerning reactions of chain side macroradicals (undergoing main-chain scission) with reactive substances added to the system.

2.2. Evaluation of Data

Two modes of information are available in principle from oscilloscope traces of the kind shown in Fig. 2: (i) The total decrease of the LSI as depicted by the difference of signal voltages $U_0 - U_\infty$ (the subscripts 0 and ∞ designate the state of the system before and a long time after irradiation, respectively). (ii) The kinetics of the change of the LSI, i.e. the function $U = \mathrm{f}(t)$.

In this section, equations are derived which allow the degree of degradation to be estimated from the total LSI decrease and kinetic parameters to be evaluated from the decrease of the LSI as a function of time (t). The considerations are based upon the fact that the scattering intensity I_s is proportional to the signal voltage U delivered by the photomultiplier.

2.2.1. *Estimation of the Degree of Degradation*

The change of the LSI during main-chain degradation of a polymer in solution may be derived from Debye's equation:

$$\frac{KcI_p}{I_s - I_L} = \frac{I}{P_\vartheta mn} + 2Ac \tag{I}$$

where

$$K = \frac{4\pi^2 \tilde{n}_0}{\lambda_0^4 N_A} \left(\frac{d\tilde{n}}{dc} \right)^2$$

c = polymer concentration in g/cm^3; I_s = intensity of light scattered by the solution; I_L = intensity of light scattered by the solvent; I_p = intensity of incident light; P_ϑ = particle scattering factor; m = molecular weight of the base unit (structural repeating unit); n = weight average degree of polymerisation; A = second virial coefficient in mol cm^3/g^2; λ_0 = wavelength of incident light; N_A = Avogadro number and $\tilde{n}$ = index of refraction.

From eqn. (I) one obtains:

$$\frac{KI_p(I_{s,0} - I_{s,\infty})}{(I_{s,0} - I_L)(I_{s,\infty} - I_L)} = \frac{1}{cm}\left(\frac{1}{(P_\vartheta)_\infty n_\infty} - \frac{1}{(P_\vartheta)_0 n_0} \right) + 2(A_\infty - A_0) \quad \text{(II)}$$

A simpler expression is derived if changes of P_ϑ and A can be neglected:

$$\frac{KI_p(I_{s,0} - I_{s,\infty})}{(I_{s,0} - I_L)(I_{s,\infty} - I_L)} = \frac{1}{cmP_\vartheta}\left(\frac{1}{n_\infty} - \frac{1}{n_0} \right) \quad \text{(III)}$$

Actually both P_ϑ and A depend upon the chain length. (Usually P_ϑ increases, if the conformation is not changed upon degradation, and A decreases with increasing chain length.) One might be permitted to consider P_ϑ and A as being constant for a small degree of degradation.

If the initial molecular weight distribution is random, $\overline{\mathrm{DP}}_{w,0} = 2\overline{\mathrm{DP}}_{n,0}$ ($\overline{\mathrm{DP}}$ is the degree of polymerisation; $\overline{\mathrm{DP}}_w \equiv n$ in the notation of this chapter). Thus, the degree of degradation, α_∞, a long time after the irradiation (where α is the number of scissions per base unit) is:

$$\alpha_\infty = \frac{2}{n_\infty} - \frac{2}{n_0} \quad \text{(IV)}$$

From eqns. (III) and (IV) it follows that:

$$\alpha_\infty = \frac{2cmP_\vartheta KI_p(I_{s,0} - I_{s,\infty})}{(I_{s,0} - I_L)(I_{s,\infty} - I_L)} \quad \text{(V)}$$

Substitution of $(I_{s,0} - I_L)$ in eqn. (V) according to eqn. (I′)

$$(I_{s,0} - I_L) = \frac{KcI_p P_\vartheta m n_0}{1 + 2Acn_0 m P_\vartheta} \quad \text{(I′)}$$

yields:

$$\frac{U_0 - U_\infty}{U_\infty - U_L} = \left(\frac{n_0}{4AcP_\vartheta mn_0 + 2}\right)\alpha_\infty \tag{VI}$$

Based on eqn. (VI), it is possible to estimate the radiation or photochemical yields of main-chain scission (at fairly low values of α_∞).

For the radiation chemical yield $G(S)$ one obtains:

$$G(S) = \frac{(U_0 - U_\infty)FN_A 10^2}{(U_\infty - U_L)mD_a} \tag{VII}$$

where $G(S)$ = number of main-chain scissions per 100 eV absorbed energy; D_a = absorbed dose in eV/g; $F = (2 + 4AcP_\vartheta mn_0)/n_0$ and U_L = signal voltage of solvent.

An expression analogous to eqn. (VII) holds for the photochemical quantum yield, $\phi(S)$:

$$\phi(S) = \frac{(U_0 - U_\infty)FN_A}{(U_\infty - U_L)mD'_a} \tag{VIII}$$

where $\phi(S)$ = number of main-chain scissions per photon and D'_a = absorbed dose in photons/g.

2.2.2. *Determination of the Lifetime, τ, of LSI*

For the derivation of equations describing the function $U = f(t)$ two possibilities must be considered, namely, that the rate of the main-chain scission process is determined by a first order or by a second order process.

(*a*) *First order kinetics.* It is assumed that the degree of degradation depends on the time t as:

$$\alpha_\infty - \alpha_t = \alpha_\infty \exp(-k_1 t) \tag{IX}$$

If eqn. (IX) is written in the form

$$\ln\frac{\alpha_\infty - \alpha_t}{\alpha_\infty} = -k_1 t \tag{X}$$

one obtains, using eqns. (V) and (I):

$$\ln\frac{(U_\infty - U_L)^{-1} - (U_t - U_L)^{-1}}{(U_\infty - U_L)^{-1} - (U_0 - U_L)^{-1}} = -k_1 t \tag{XI}$$

Equation (XI) is identical with eqn. (XI′)

$$\ln \frac{(U_0 - U_L)(U_t - U_\infty)}{(U_t - U_L)(U_0 - U_\infty)} = -k_1 t \qquad \text{(XI}')$$

$$\tau(\text{LSI}) \equiv k_1^{-1}$$

(*b*) *Second order kinetics.* It is assumed that the degree of degradation depends on the time t as:

$$\frac{1}{\alpha_\infty - \alpha_t} - \frac{1}{\alpha_\infty} = \frac{2 \times 10^3 c}{m} k_2 t \qquad \text{(XII)}$$

Substitution of α_∞ and α_t according to eqns. (V) and (I) leads to:

$$\frac{(U_\infty - U_L)(U_t - U_L)}{(U_0 - U_L)(U_t - U_\infty)} - \frac{(U_\infty - U_L)}{(U_0 - U_\infty)} = \frac{2 \times 10^3 cF}{m} k_2 t \qquad \text{(XII}')$$

3. INSTRUMENTAL TECHNIQUE

The experimental device used for the experiments described in the following sections is schematically represented in Fig. 4.

The sample cell containing the polymer solution was irradiated with a flash of u.v. light or an impulse of high energy electrons. u.v. Light flashes of $\lambda = 347{\cdot}1$ or 265 nm (flash duration of 20 ns) were produced either by a ruby laser (Korad, model K1QS2) with the aid of a frequency doubler (ammonium dihydrogen phosphate (ADP); $\lambda = 347{\cdot}1$ nm) or by a neodymium YAG laser. In this case the pulse was produced by a Nd-YAG oscillator (J. K. Lasers Ltd) and amplified by a Nd glass amplifier (Korad). The 1060 nm light was frequency-quadrupled using a KD*P crystal (potassium dideuterophosphate and an ADP crystal). Thus 265 nm light was obtained. A linear accelerator (L-band, Vickers) was used to generate impulses of fast electrons (16 MeV, pulse duration 0·050–2 μs).

As analysing light sources, argon-ion lasers were used (Spectra Physics Model 165-00). The nominal output was 2 and 4 W. The 514·5 nm line was isolated by a prism and passed through the sample cell at right angles with respect to the fast electron or u.v. light beam. For measuring the LSI at an angle $\vartheta = 90°$ the light scattered by the polymer solution was directed to a photomultiplier (EMI, 9781B) enclosed in a radio-frequency-tight box via a light guide. The latter consisted, in the case of fast electron irradiations, of a quartz rod connecting the sample cell with the photomultiplier (protected by a lead shield). In the case of u.v. irradiations commercially available

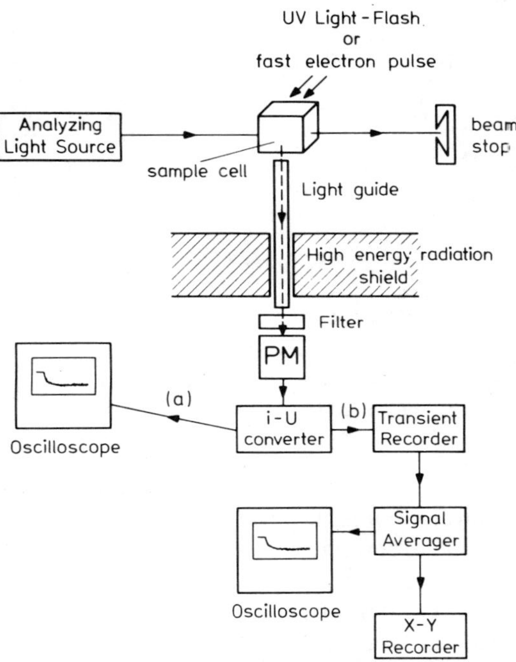

FIG. 4. Schematic representation of an apparatus for time-resolved LSI measurements.[1]

flexible light guides were used. To prevent overloading of the photomultiplier by light emission of the sample (luminescence or Čerenkov light) a nm interference filter (Oriel) was placed between the light guide and the photomultiplier. The photomultiplier was operated with six stages. A suitable dynode chain allowed linear operation up to an anode current of 2 mA. The minimum fatigue was at 200 μA. The minimum time resolution amounted to about 1 μs.

Two modes of signal display were utilised, as indicated in Fig. 4 by routes (a) and (b). In both cases the output of the photomultiplier was fed to a current-to-voltage converter with a rise time of less than 100 ns. In route (a) the signal was fed into a differential amplifier (plug-in 7A22 or 1A7A, Tektronix) which permitted the selection of the band width for optimum signal-to-noise ratio, and displayed on a storage oscilloscope (7633 or 549, Tektronix). In route (b) the signal was digitised by a transient recorder (model 8100, Biomation) and transferred to the memory of a signal averager (model 1072 instrument computer, Nicolet). An appropriate

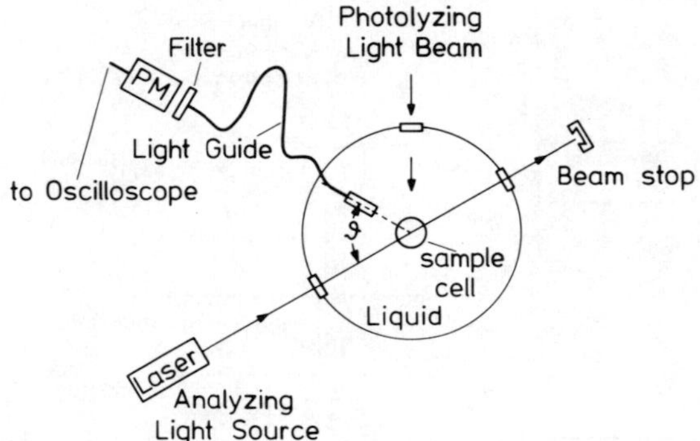

FIG. 5. Schematic representation of an apparatus for time-resolved LSI measurements at different angles.[5]

number of LSI signals (usually 4 or 16) and a corresponding number of light emission signals were stored separately in the memory of the averager. Subtraction yielded the corrected LSI signal. Route (b) was utilised only in the case of fast electron irradiations where the pulse triggering reproducibility was of sufficient accuracy.

For the determination of LSI changes at various scattering angles the device represented schematically in Fig. 5 was used. It consisted of a metal vat filled with an appropriate liquid. A cylindrical quartz cuvette with the polymer solution was placed in the centre of the vat.

The vat had three windows, two of which provided for the passage of the analysing light beam and the third served as entrance window for the photolysing light. This device was used exclusively for photolysis studies. The vat was covered by a round, rotary plate which contained at fixed positions the holder for the sample cell and a supporter for the flexible light guide. This supporter kept one end of the light guide at a constant distance from the cuvette. The other end was fixed to the photomultiplier. By rotation of the round plate the entrance face of the light guide could be placed at various positions pertaining to scattering angles from 30 to 150°.

4. DYNAMIC STUDIES

4.1. Coiled Macromolecules

4.1.1. *Disentanglement Diffusion*

During the investigation of numerous polymers two compounds were found to be appropriate for studying disentanglement diffusion, namely polyisobutene (PIB) and poly(phenylvinylketone) (PPVK):

$$-\!\!\left[CH_2-C(CH_3)_2 \right]_n\!\!- \quad \text{and} \quad -\!\!\left[CH(C{=}O\,C_6H_5)-CH_2 \right]_n\!\!-$$

(PIB) (PPVK)

(*a*) *Pulse radiolysis studies on polyisobutene* (*PIB*). Polyisobutene dissolved in saturated hydrocarbons was degraded by irradiation with 2 μs impulses of 16 MeV electrons.[2] Typical oscilloscope traces and corresponding plots according to eqn. (XI′) are shown in Fig. 6. The half-life time, $\tau_{1/2}$(LSI), was not influenced by the addition of cyclohexane, which, on the other hand, drastically reduced the extent of degradation.

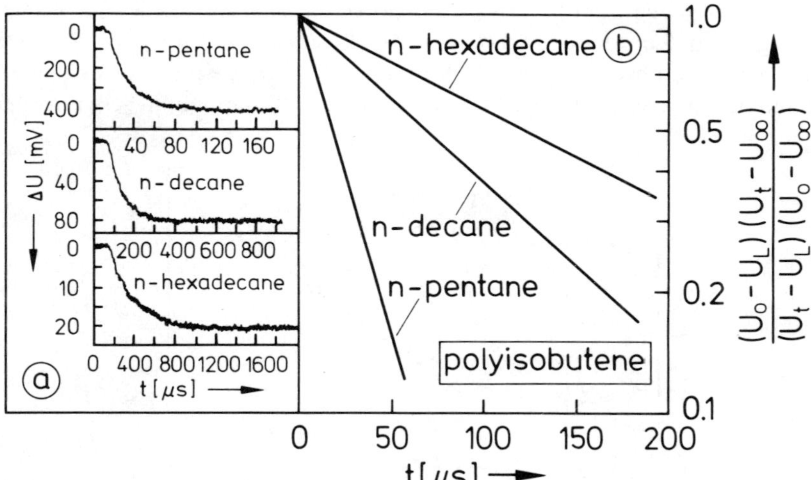

FIG. 6. Degradation of polyisobutene in hydrocarbon solutions at room temperature. (a) Oscilloscope traces obtained by monitoring the change of LSI in different solvents after irradiation with a pulse of 16 MeV electrons. (b) First order plot according to eqn. (XI′). $\bar{M}_{w,0} = 2{\cdot}1 \times 10^6$; $\bar{M}_{n,0} = 1{\cdot}3 \times 10^6$; $[PIB] = 2{\cdot}7 \times 10^{-2}$ base mol/litre; absorbed dose 4×10^4 rad; pulse length = 2 μs.[2]

TABLE 1

DEGRADATION OF POLYISOBUTENE. INFLUENCE OF CYCLOHEXENE ON THE EXTENT OF DEGRADATION AND THE HALF-LIFE TIME OF LSI DECREASE. $\bar{M}_{w,0} = 2{\cdot}1 \times 10^6$; $\bar{M}_{n,0} = 1{\cdot}3 \times 10^6$. [PIB] = 0·018 BASE MOL/LITRE; ABSORBED DOSE: *ca.* $2{\cdot}5 \times 10^4$ RAD.[2]

Vol % *cyclohexane*	$\frac{U_0 - U_\infty}{U_\infty - U_L}$	$\tau_{1/2}$(*LSI*) (μs)
0	0·55	23·0 ± 1·5
1	0·26	24·0 ± 1·5
2	0·25	25·0 ± 1·5
5	0·16	26·0 ± 1·5
10	0·14	26·0 ± 1·5

This is demonstrated by the results in Table 1. It is therefore concluded that the time for establishing bond scissions in the main chain is much shorter than $\tau_{1/2}$(LSI) and that the latter corresponds to disentanglement diffusion. The extent of degradation was proportional to the absorbed dose as shown in Fig. 7, in accordance with eqn. (VII). The half-life time (LSI) did not depend on the absorbed dose in the investigated range which corresponded to maximum average values of Z_s (the number of main-chain scissions per

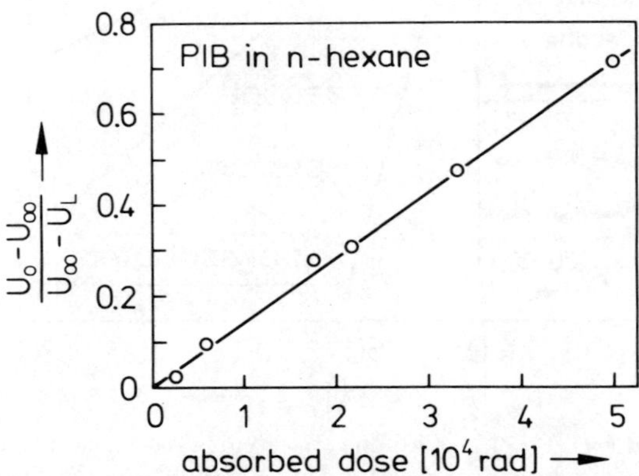

FIG. 7. Degradation of polyisobutene in n-hexane at room temperature. Dependence of the extent of main-chain scission on the absorbed dose. Plot according to eqn. (VII). $\bar{M}_{w,0} = 2{\cdot}6 \times 10^5$; $\bar{M}_{n,0} = 2{\cdot}1 \times 10^5$; [PIB] = $1{\cdot}8 \times 10^{-2}$ base mol/litre.[2]

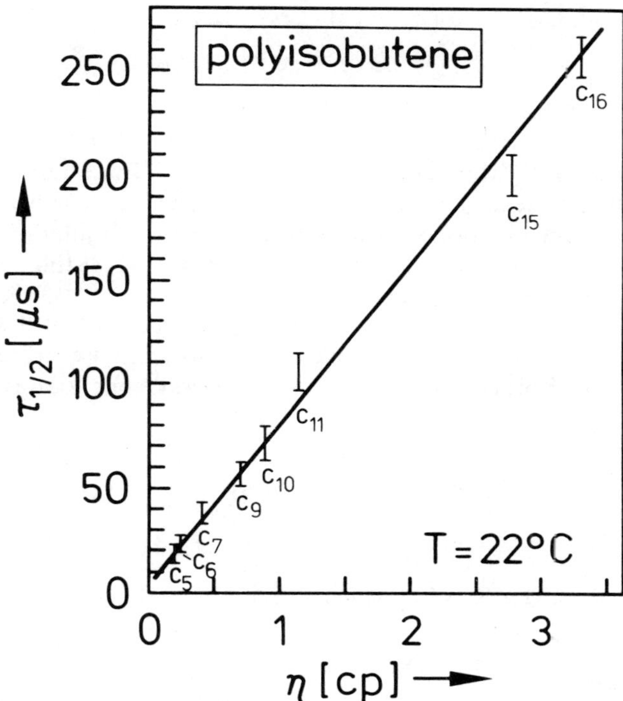

FIG. 8. Degradation of polyisobutene in hydrocarbons (n-pentane to n-hexadecane) at room temperature. Influence of the microviscosity on the half-life time $\tau_{1/2}$(LSI). $\bar{M}_{w,0} = 2{\cdot}1 \times 10^6$; $\bar{M}_{n,0} = 1{\cdot}3 \times 10^6$; [PIB] $= 2{\cdot}7 \times 10^{-2}$ base mol/litre; absorbed dose $= 4 \times 10^4$ rad.[2]

initial macromolecule) of about 1·3. It was found that $\tau_{1/2}$(LSI) was directly proportional to the microviscosity (solvent viscosity) as expected if $\tau_{1/2}$(LSI) corresponds to a diffusion process. The results are presented in Fig. 8. Furthermore, it was found that $\tau_{1/2}$(LSI) $\propto M_0^{0{\cdot}34}$. If $\tau_{1/2}$(LSI) were to be correlated with a pure translational diffusion process one would expect a much stronger dependence of $\tau_{1/2}$(LSI) on M_0, e.g. $\tau_{\text{trans.}} \propto M^{1{\cdot}5}$ (M = molecular weight). This proportionality is estimated by assuming that the macromolecules are split into fragments of equal end-to-end distance R. Then $\tau_{\text{trans.}}$ depends on R and the translational diffusion constant $D_{\text{trans.}}$ as in:

$$\tau_{\text{trans.}} \propto \frac{R^2}{D_{\text{trans.}}} \qquad \text{(XIII)}$$

From eqn. (XIII) it follows that:

$$\tau_{\text{trans.}} \propto M^{2b+c} \qquad \text{(XIII}')$$

since $R \propto M^b$ and $D_{\text{trans.}} \propto M^{-c}$.

For macromolecules with coil conformation b is 0·5–0·7 and c is *ca*. 0·5. Therefore, eqn. (XIII′) yields $\tau_{\text{trans.}} \propto M^{1\cdot5}$–$M^{1\cdot9}$. These results together with studies of the dependence of $\tau_{1\,2}$(LSI) on the coil density (*vide infra*) indicate that the rate of LSI change observed with PIB solutions is due to disentanglement diffusion and that the latter is to be distinguished from pure translational diffusion.

(*b*) *Flash photolysis studies on poly*(*phenylvinylketone*) (*PPVK*). PPVK undergoes main-chain degradation on irradiation with u.v. light according to a Norrish II process involving triplet excited ketone groups:

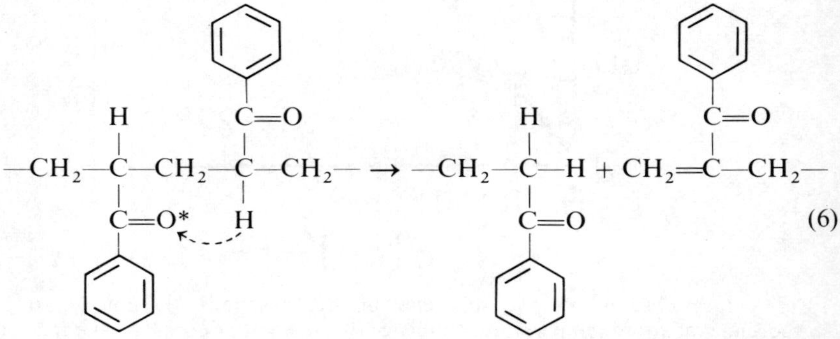

$$\text{—CH}_2\text{—C(H)(C{=}O^{*}\text{—C}_6\text{H}_5)\text{—CH}_2\text{—C(H)(C{=}O\text{—C}_6\text{H}_5)\text{—CH}_2\text{—} \rightarrow \text{—CH}_2\text{—C(H)(C{=}O\text{—C}_6\text{H}_5)\text{—H} + \text{CH}_2\text{=C(C{=}O\text{—C}_6\text{H}_5)\text{—CH}_2\text{—} \qquad (6)$$

Reaction (6) occurs at times shorter than 100 ns, as determined by triplet lifetime measurements.[3] A very short-lived biradical might be formed as an intermediate:

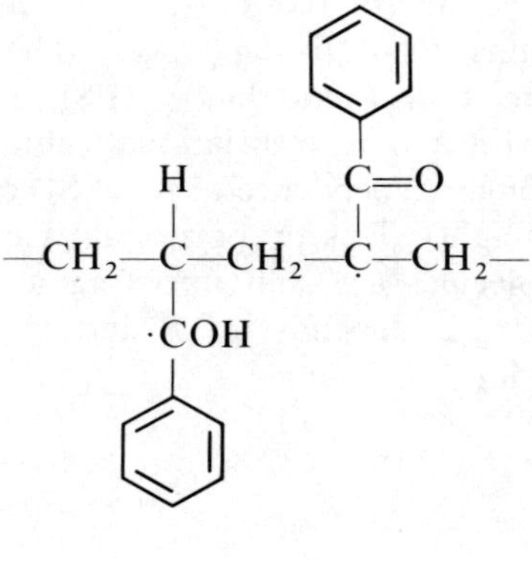

$$\text{—CH}_2\text{—C(H)(}\cdot\text{COH—C}_6\text{H}_5)\text{—CH}_2\text{—}\dot{\text{C}}\text{(C{=}O—C}_6\text{H}_5)\text{—CH}_2\text{—}$$

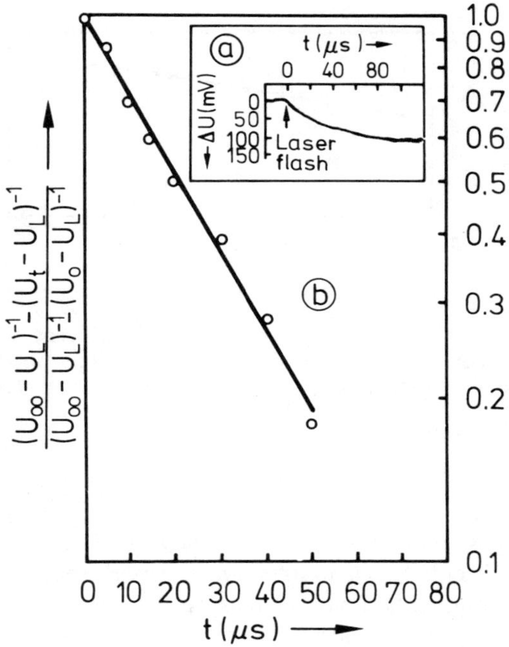

FIG. 9. Degradation of PPVK in benzene solution at room temperature. (a) Oscilloscope trace demonstrating the time dependence of the LSI decrease after a 25 ns flash of 346·1 nm light. $\vartheta = 90°$; [PPVK] = $3{\cdot}8 \times 10^{-3}$ base mol/litre; $\bar{M}_{w,0} = 4{\cdot}6 \times 10^5$. (b) Plot according to eqn. (XI) of data from Fig. 9(a).[1]

After the irradiation of dilute PPVK solutions with 25 ns flashes of 347·1 nm light the LSI decayed with $\tau_{1/2}$(LSI) > 10 μs. Therefore, it is obvious that the chemical step of bond scission occurs in this case several orders of magnitude faster than the separation of scission fragments. A typical oscilloscope trace is shown in Fig. 9. It may be emphasised here that PPVK served as an excellent probe for studying disentanglement diffusion processes. The photolytic main-chain rupture process is relatively well understood. Furthermore, it was possible to vary Z_s, the number of main chain breaks per initial number average molecular weight molecule produced during a single flash, over a very broad range. Thus, the influence of fragment size on the rate of disentanglement diffusion could be easily studied. Early investigations[1] with PPVK in benzene solution (carried out under conditions with $Z_s \leq 1$) showed (Fig. 10) that the influence of the initial average molecular weight upon the rate of fragment separation is

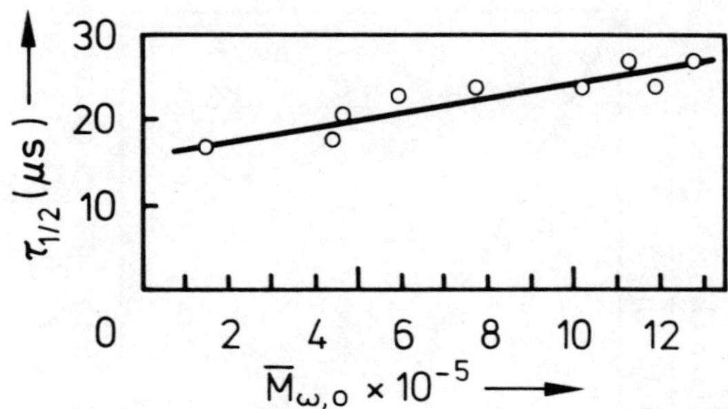

FIG. 10. Degradation of poly(phenylvinylketone) in benzene solution at room temperature. $\tau_{1/2}$(LSI) as a function of the initial molecular weight. [PPVK] = $3{\cdot}8 \times 10^{-3}$ base mol/litre; irradiation with 25 ns flashes of 347 nm light.[1]

rather weak ($\tau_{1/2} \propto \bar{M}_{w,0}^{0{\cdot}22}$) in analogy to the results obtained with hydrocarbon solutions of PIB. With PPVK, $\tau_{1/2}$(LSI) was also found to increase linearly with increasing solvent viscosity. Thus, it was concluded that in this case disentanglement diffusion is not determined by translatory diffusion of the fragments. Upon increasing Z_s it was found that $\tau_{1/2}$(LSI) was constant up to a critical value Z_s(crit) ≈ 10, as shown in Fig. 11.[4] These experiments were carried out with dioxane solutions. (Dioxane has a relatively high viscosity, 1·44 cP at 20 °C, so that $\tau_{1/2}$(LSI) values are appreciably higher than for benzene solution, where the intrinsic viscosity $\eta_{(\text{benzene})} = 0{\cdot}65$ cP at 20 °C.)

It is obvious from Fig. 11 that at $Z_s > Z_s$(crit), $\tau_{1/2}$(LSI) decreases

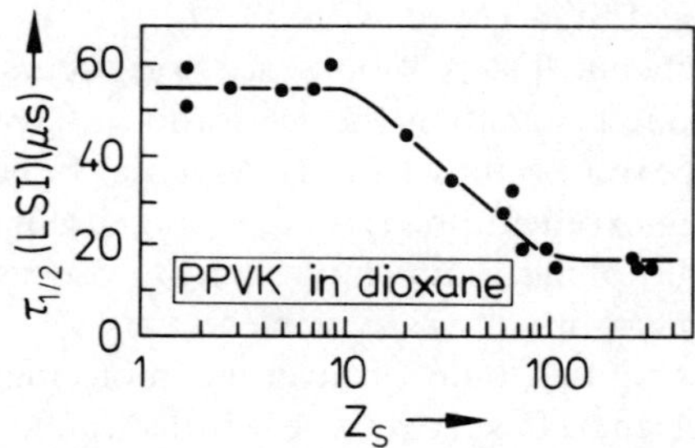

FIG. 11. Degradation of poly(phenylvinylketone) in dioxane solution at room temperature. $\tau_{1/2}$(LSI) versus the number of scissions per initial macromolecule. $\bar{M}_{w,0} = 1{\cdot}5 \times 10^6$; [PPVK] = $3{\cdot}8 \times 10^{-3}$ base mol/litre; irradiation with 25 ns flashes of 347 nm light.[4]

significantly with increasing values of Z_s. Obviously, the mechanism of disentanglement separation changes with variation of the fragment size, thus as the fragment size decreases and passes a certain critical size the rate of fragment separation becomes faster and probably becomes determined by the rate of translational motion of the fragments. This effect can be understood on the basis of intramolecular contact-pair formation which exerts a significant influence on the disentanglement diffusion in the case of long chains (big fragment size). This influence diminishes with decreasing chain length and becomes almost completely unimportant at low values of the chain length.

For a given polymer/solvent system the probability w_{CP} for contact-pair formation is determined by a thermodynamic interaction parameter and is considered to be constant under fixed conditions, e.g. of temperature, pressure, etc. (w_{CP} is defined as the probability that a single base unit of a chain approaches another base unit of the same chain so closely that physical interaction (dipole–dipole, Van der Waals, etc.) becomes possible). The lifetime of a contact pair is significantly longer than the encounter time. A simple model might be based on the assumption that w_{CP} is independent of the degree of polymerisation (DP). Then, the number of contact pairs per macromolecule is given by eqn. (XIV)

$$n_{CP} = \tfrac{1}{2} w_{CP} (\mathrm{DP})^2 \qquad \text{(XIV)}$$

An influence of contact pairs on disentanglement diffusion is expected only if DP exceeds a critical chain length $(\mathrm{DP})_{crit}$, i.e. if n_{CP} is significantly greater than unity. In other words if the fragment size is rather small ($(\mathrm{DP}) < (\mathrm{DP})_{crit}$), the separation of fragments is not influenced by intramolecular interaction.

The concept of contact-pair formation being responsible for the fact that intramolecular interaction is determining the rate of disentanglement at large values of the chain length, has been supported by a series of experiments carried out with PPVK samples of different initial molecular weights at various scattering angles.[5] As shown in Fig. 12, $\tau_{1/2}$(LSI) decreases at all scattering angles between 45 and 90° with increasing values of Z_s for $\bar{M}_{w,0} = 1{\cdot}15 \times 10^5$. For the PPVK samples of higher molecular weight more or less broad regions exist where $\tau_{1/2}$(LSI) is independent of Z_s. Obviously there is a trend for Z_s(crit) to shift to lower values of Z_s as the scattering angle ϑ becomes greater. The observed effects can be explained by the fact that $(\mathrm{DP})_{crit}$ corresponds to about 10^3 for the system PPVK/dioxane. If the initial DP is smaller than $(\mathrm{DP})_{crit}$, intramolecular interaction will not influence the rate of disentanglement diffusion.

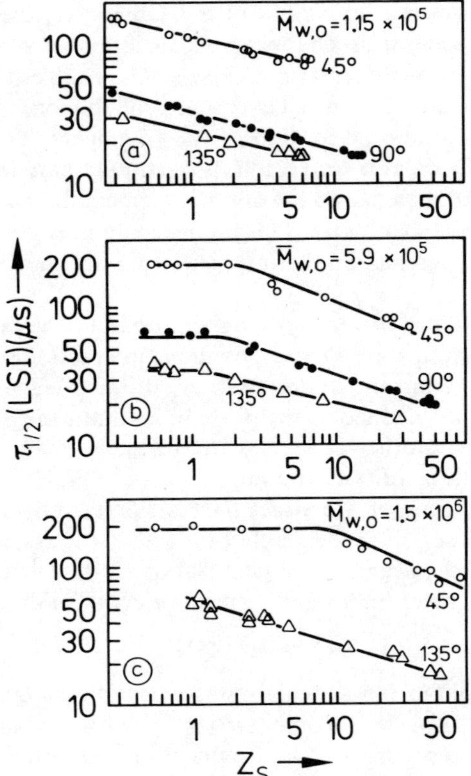

FIG. 12. Degradation of poly(phenylvinylketone) in dioxane solution at room temperature. $\tau_{1/2}$(LSI) as a function of the number of scissions per initial macromolecule at various scattering angles. [PPVK] = 5×10^{-3} base mol/litre. Irradiation with 25 ns flashes of 347 nm light. Initial weight average molecular weight = (a) $1{\cdot}15 \times 10^5$, (b) $5{\cdot}9 \times 10^5$, and (c) $1{\cdot}5 \times 10^6$.[5]

Differences in the function $\tau_{1/2}$(LSI) = f(Z_s) found at various angles can be explained as follows: Since photolytically induced main-chain scissions in PPVK are distributed at random along the chains, the size of the fragments also corresponds to a random distribution. In other words, at low values of Z_s some macromolecules are fragmented into small and large portions; others are ruptured into fragments of similar size. The size distribution of fragments will cause differences in the rate of LSI change observed at various scattering angles. The large particles contribute more strongly to the LSI at low angles than the small ones, since large particles scatter more strongly in the forward than in the reverse direction. Thus, it becomes understandable that at constant Z_s, $\tau_{1/2}$(LSI) decreases with increasing scattering angle and that intramolecular interaction becomes detectable with the high molecular weight samples the lower the scattering angle. Both effects are inferred from the results presented in Fig. 12.

If intramolecular contact-pairs exert an influence on disentanglement diffusion, the solvent quality should also play a pronounced role. As the

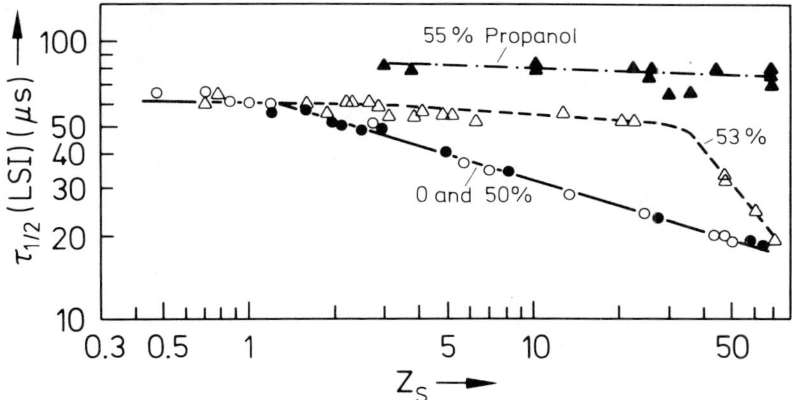

FIG. 13. Degradation of poly(phenylvinylketone) in dioxane/n-propanol mixtures at room temperatures. $\tau_{1/2}$(LSI) versus the number of main-chain scissions per initial macromolecule. $\bar{M}_{w,0} = 5{\cdot}9 \times 10^5$; [PPVK] $= 5 \times 10^{-3}$ base mol/litre; irradiation with 25 ns flashes of 347 nm light.[8]

solvent becomes less good the number of contact pairs increases, since the probability w_{CP} is increasing (at constant DP). Recently it was reported[6] that at $T = \theta$:

$$n_{CP} \propto (\mathrm{DP})^{0{\cdot}5} \qquad \text{(XV)}$$

and for a good solvent ($T \gg \theta$)

$$n_{CP} \propto \ln(\mathrm{DP}) \qquad \text{(XVI)}$$

At constant DP, n_{CP} should, therefore, be significantly higher in a θ solvent than in a good solvent. As the solvent becomes less good the critical fragment size above which disentanglement diffusion is determined by intramolecular interaction, should be shifted to lower values. This expectation is corroborated by the results shown in Fig. 13, where again $\tau_{1/2}$(LSI) is plotted versus Z_s.[8] In this case the solvent consisted of various mixtures of dioxane and n-propanol (viscosities at 20 °C: *p*-dioxane: 1·44 cP; n-propanol: 2·26 cP). The precipitation point at 23 °C corresponds to a n-propanol/dioxane composition of 57/43 (vol/vol). If a precipitant is added to a solvent the coil density of a dissolved polymer changes significantly upon approaching the precipitation point. This has been demonstrated, for example, for the system PIB/n-heptane/n-propanol by Kunst.[7] As shown in Fig. 14, in the case of PIB, $\tau_{1/2}$(LSI) increased

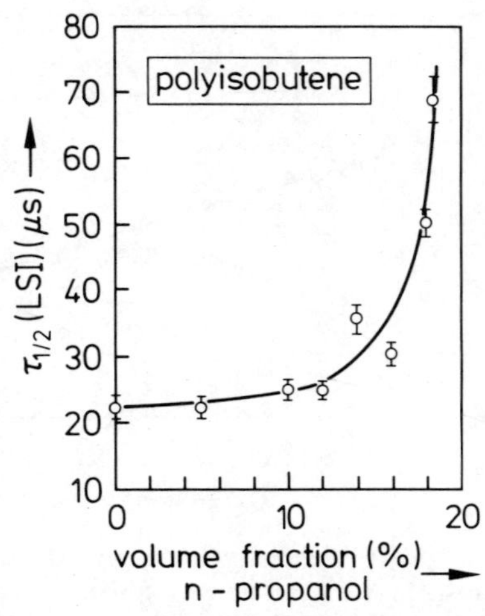

FIG. 14. Degradation of polyisobutene in n-hexane/n-propanol mixtures at room temperature. $\tau_{1/2}$(LSI) versus the volume fraction of n-propanol. $\bar{M}_{w,0} = 2{\cdot}1 \times 10^6$; [PIB] $= 2{\cdot}7 \times 10^{-2}$ base mol/litre; absorbed dose per pulse $= 4 \times 10^4$ rad; irradiation with 2 μs impulses of 16 MeV electrons.[2]

drastically at solvent mixtures corresponding to compositions in the neighbourhood of the precipitation point of about 20 vol % n-propanol at 23 °C. Concerning the results in Fig. 13, an analogous behaviour was found for the system PPVK/dioxane/n-propanol. At 50 vol % n-propanol the same dependence of $\tau_{1/2}$(LSI) on Z_s was found as at n-propanol contents of 0 and 50 vol %. At 53 and 55 vol % n-propanol, however, a strongly pronounced effect was observed. At a n-propanol content of 53 vol %, Z_s(crit) was shifted to about 30. At 55 vol %, Z_s(crit) was not reached in the investigated region of Z_s, indicating the high number of intramolecular contact-pairs existing under these conditions.

Preliminary experiments with copolymers of phenylvinylketone (PVK) and styrene or methylmethacrylate containing small portions of PVK have shown that the rate of LSI decrease corresponds in these cases also to disentanglement diffusion.[9] With the aid of copolymers containing a small portion of easily degradable base units it is, therefore, possible to study disentanglement diffusion of polymers such as poly(methylmethacrylate) or polystyrene, which have served frequently as objects for thermodynamic studies.

Concluding this section, it may be pointed out that time-resolved light-scattering measurements in combination with a powerful and rapid mode of main-chain degradation have been utilised in order to investigate disentanglement diffusion. These results might also be valuable for the understanding of inverse processes, namely the entanglement or partial overlapping of coils which occurs, for example, during the combination of macroradicals, a case which is also met during the kinetic treatment of polymerisation processes.

4.1.2. *Preferential Solvation*

During the study of disentanglement diffusion in binary solvent mixtures, results were obtained which can be associated with the phenomenon of preferential solvation.[9,10] When poly(methylmethacrylate-*co*-phenylvinylketone) containing 7·8 mol % PVK was flash irradiated in acetone solution, $\tau_{1/2}$(LSI) was found to increase upon the addition of naphthalene as shown in Fig. 15. At naphthalene concentrations greater than 0·05 mol/litre $\tau_{1/2}$(LSI) approaches a limiting value. As shown by the decrease of the degree of degradation with increasing naphthalene concentration,

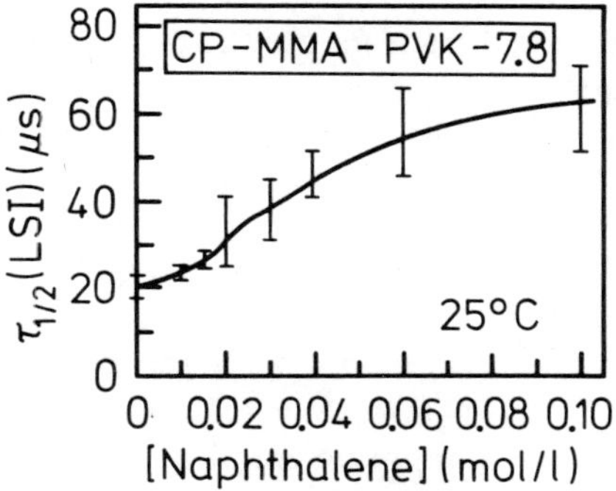

FIG. 15. Degradation of poly(methylmethacrylate-*co*-phenylvinylketone) containing 7·8 mol % PVK at 25 °C. $\tau_{1/2}$(LSI) versus naphthalene concentration. [Polymer] = $2{\cdot}1 \times 10^{-2}$ base mol/litre; irradiation with 25 ns flashes of 347 nm light; absorbed dose per flash = $8{\cdot}4 \times 10^{-5}$ Einstein/litre.[10]

naphthalene and 2,5-dimethyl-2,4-hexadiene act as triplet quenchers. The latter compound did not exert an influence on $\tau_{1/2}$(LSI). (If $\tau_{1/2}$(LSI) were related to a chemical reaction it should decrease with increasing triplet quencher concentration.) The results can be understood on the basis of preferential solvation as depicted schematically in Fig. 16.

In pure acetone the macromolecules are solvated by acetone molecules. As naphthalene is added, successively more and more solvating acetone

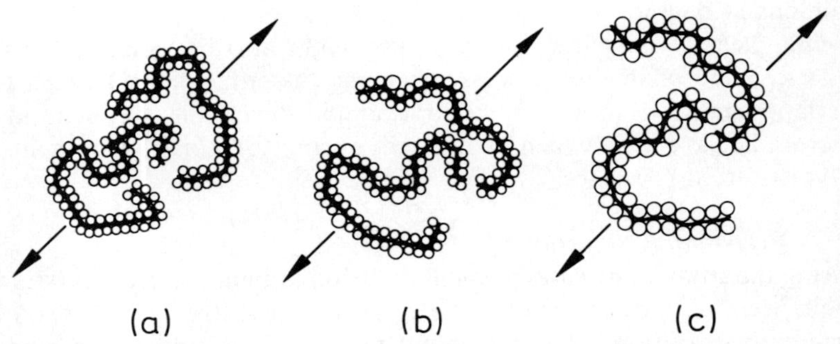

FIG. 16. Schematic representation of the influence of preferential solvation on the separation of fragments. (a) Totally solvated by small molecules, (b) partially solvated by bulky molecules, (c) totally solvated by bulky molecules.

molecules are replaced by bulky naphthalene molecules. Thus, the mobility of the macromolecules is decreased, that is, the time of fragment separation is increased. The fact that $\tau_{1/2}$(LSI) approaches a limiting value confirms this assumption. Additional evidence for the preferential solvation of the copolymer was obtained by measuring the intrinsic viscosity $[\eta]$. On addition of 0·1 mol/litre naphthalene, $[\eta]$ increased by a factor of 1·13. An increase of $[\eta]$ indicates coil expansion or a decrease of n_{CP} (the number of intramolecular contact-pairs). According to the contact-pair concept (*vide ante*) the addition of naphthalene should, therefore, lead to higher rates of fragment separation, that is, to lower values of $\tau_{1/2}$(LSI). The inverse behaviour was found, in accordance with the concept of preferential solvation.

Further work[11] concerning preferential solvation was devoted to systems containing poly(vinylpyrrolidone) (solvent: $CHCl_3$/2-naphthol or $CHCl_3$/phenol). Preferential adsorption in these systems has also been studied (by other means) by Chaufer *et al.*[12] The results are shown in Fig. 17. On addition of 2-naphthol or phenol to a solution of

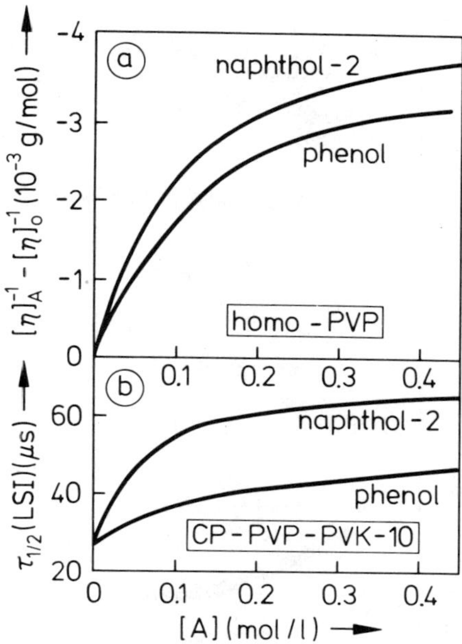

FIG. 17. Influence of preferential solvation on the intrinsic viscosity (a) and the rate of fragment separation (b) in chloroform solution at room temperature. (a) $[\eta]_{add.}^{-1} - [\eta]_0^{-1}$ versus the concentration of additive. Poly(N-vinylpyrrolidone): $\bar{M}_{w,0} = 1{\cdot}1 \times 10^6$, (b) $\tau_{1/2}$(LSI) versus the concentration of additive. Poly(N-vinylpyrrolidone-*co*-phenylvinylketone) containing 10 mol % PVK. $\bar{M}_{n,0} = 4{\cdot}4 \times 10^5$. [Polymer] = 4×10^{-2} base mol/litre.[11]

poly(N-vinylpyrrolidone-*co*-phenylvinylketone) containing 10 mol % PVK, $\tau_{1/2}$(LSI) increased and approached a limiting value (by analogy to the results obtained with the system copolymer–MMA–PVK–7·8/acetone/naphthalene). This is shown in Fig. 17(b). The change of $\tau_{1/2}$(LSI) is paralleled by an increase of $[\eta]$. In Fig. 17(a) the difference $[\eta]_{add.}^{-1} - [\eta]_0^{-1}$ as measured for a homo-PVP sample is plotted versus the concentration of the additive. This difference is frequently used to measure the extent of preferential solvation (sometimes defined as the coefficient of preferential solvation).[13,14] The negative sign of the difference $[\eta]_{add.}^{-1} - [\eta]_0^{-1}$ indicates that the additive is preferentially adsorbed by the polymer.

4.2. Deoxyribonucleic Acid (DNA)

4.2.1. *Native DNA*

Results obtained upon irradiation of aqueous solutions of native calf-

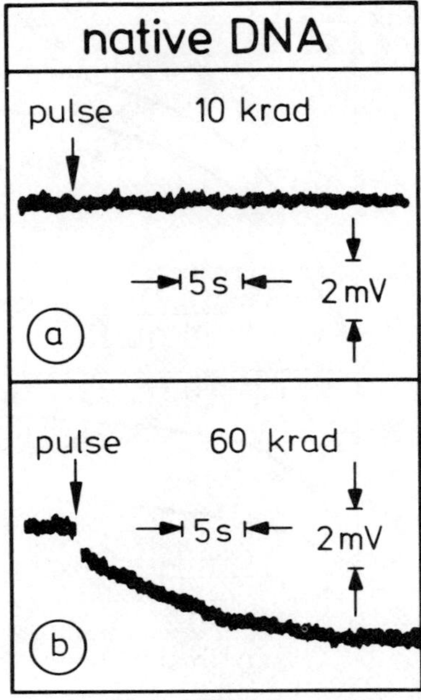

FIG. 18. Degradation of native calf-thymus DNA. Oscilloscope traces demonstrating the change of the LSI after irradiation with an impulse of 16 MeV electrons at room temperature. [DNA] = $7{\cdot}5 \times 10^{-2}$ g/litre. $U_0 = 23$ mV; absorbed dose = (a) 1×10^4 rad, (b) 6×10^4 rad.[15]

thymus DNA with 2 μs pulses of 16 MeV electrons strongly reflect the double strand nature of this biopolymer.[15,16] Typical oscilloscope traces obtained at room temperature are shown in Fig. 18. At a relatively low absorbed dose of *ca.* 1×10^4 rad the LSI was unchanged. At $D_a >$ $1{\cdot}3 \times 10^4$ rad a relatively slow decrease of the LSI was observed and at $D_a > 4 \times 10^4$ rad, two modes of LSI change after the pulse were detected. As depicted in Fig. 19 the fast decrease ($\tau_{1/2}$(LSI) = 1–30 ms, depending on dose, *vide infra*) is ascribed to the separation of DNA fragments produced by double strand breaks situated directly opposite to each other (Fig. 19(a)). The slow decrease ($\tau_{1/2}$(LSI) $\approx$ 8 s) is attributed to the detachment of segments generated by single strand breaks at sites on alternate strands being separated by several nucleotide units, as depicted in Fig. 19(b). (An estimate yielded an average segment length between the breakpoints of about 10 nucleotide units.) As indicated by the results shown in Fig. 18, LSI changes can only be detected if the absorbed dose exceeds a critical value.

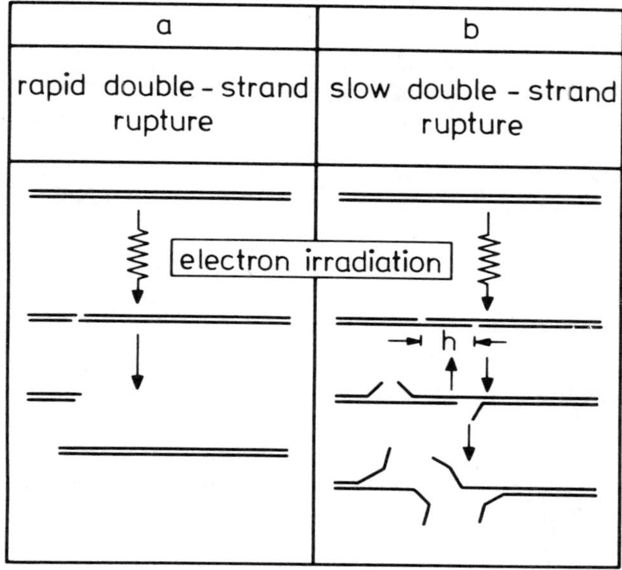

FIG. 19. Schematic representation of the generation of double strand breaks in DNA.[15]

This is due to the fact that main-chain scissions are distributed at random along the chains. The degree of degradation being solely due to double strand breaks was found to increase exponentially with absorbed dose in our case, in accordance with earlier findings obtained during stationary irradiations with γ-rays.[17] At a low dose per pulse the probability of double strand breaks was too low for degradation to become detectable. A change in LSI could then be observed only if a second or third pulse was applied. Table 2 shows that $\tau_{1/2}(\text{LSI})_{\text{fast}}$ is about 30 ms on irradiating the DNA solution with a single pulse of 5×10^4 rad. It decreased to about 1 ms at 21 °C when a second pulse was applied within *ca.* 60 s. This is due to the fact that with increasing dose the average fragment length decreases and that the shorter the fragment size the faster the rate of separation. Furthermore, $\tau_{1/2}(\text{LSI})_{\text{fast}}$ decreased with decreasing initial molecular weight ($\tau_{1/2}(\text{LSI})_{\text{fast}} = 40$ ms at $\bar{M}_{w,0} = 1{\cdot}2 \times 10^7$, and $\tau_{1/2}(\text{LSI})_{\text{fast}} = 10$ ms at $\bar{M}_{w,0} = 9 \times 10^5$). This decrease is due to the fact that the average fragment size is proportional to the initial chain length and that the rate of fragment separation is proportional to the reciprocal of the fragment size. The dependence of $\tau_{1/2}(\text{LSI})_{\text{fast}}$ on the absorbed dose and the initial molecular

TABLE 2

DEPENDENCE OF $\tau_{1/2}(\text{LSI})_{\text{fast}}$ ON THE ABSORBED DOSE. NATIVE DNA ($\bar{M}_{w,0} = 1{\cdot}2 \times 10^7$) IN OXYGEN SATURATED 0·01 M NaCl SOLUTION; [DNA] = 8×10^{-2} G/LITRE.[16]

Number of pulses	*Absorbed dose per pulse (10^4 rad)*	*Total absorbed dose (10^4 rad)*	T (°C)	$\tau_{1/2}(LSI)_{fast}$ (ms)
1	5	5	23	30
2	5	10	21	0·9

weight provides evidence for the fact that the rate of the fast LSI change is associated with a diffusion process. On the other hand, $\tau_{1/2}(\text{LSI})_{\text{slow}}$ did not depend on the absorbed dose, indicating that it does not depend on the diffusion of the fragments, but rather on the breaking of the hydrogen bonds between two single strand breaks on alternate strands and the subsequent detachment of the segments. The degree of degradation and the rate of LSI decay were found to depend strongly on temperature.[16] In this section temperature effects shall be discussed in some detail. The oscilloscope traces shown in Fig. 20 indicate that at temperatures below room temperature the LSI decrease is essentially due to the slow process, whereas at higher temperatures both a fast and a slow mode of LSI decay are observed. Above 40 °C the fast process becomes predominant. The increase in the proportion of fast LSI decay with increasing temperature is shown in Fig. 21. Evidence was, furthermore, obtained for the fact that above 40 °C three distinct modes of LSI decrease can be distinguished. The temperature dependence of the half-life times of the two predominant portions is shown in Fig. 22. $\tau_{1/2}(\text{LSI})_{\text{slow}}$ decreases from 10 s to 10 ms (i.e. by a factor of 10^3), whereas $\tau_{1/2}(\text{LSI})_{\text{fast}}$ is almost constant up to 40 °C and decreases significantly only at higher temperatures. As previously mentioned, three modes of LSI decrease could be detected at rather high temperatures. At 58 °C the three modes correspond to the following portions: very fast ($\tau_{1/2} \approx 10^{-3}$ s) 40 %, fast ($\tau_{1/2} \approx$ 2–5 ms) 20 % and slow ($\tau_{1/2} \approx$ 6–10 s) 20 %.

The temperature effects can be explained with the aid of Table 3 as follows: All double strand breaks that contribute to the rapid process are due to single strand breaks separated by h_f nucleotide units. On the other hand, there exists a critical number of nucleotide units h_{crit} above which single strand breaks will not become effective as double strand breaks. As depicted in Table 3, it can be assumed that the increase in the degree of

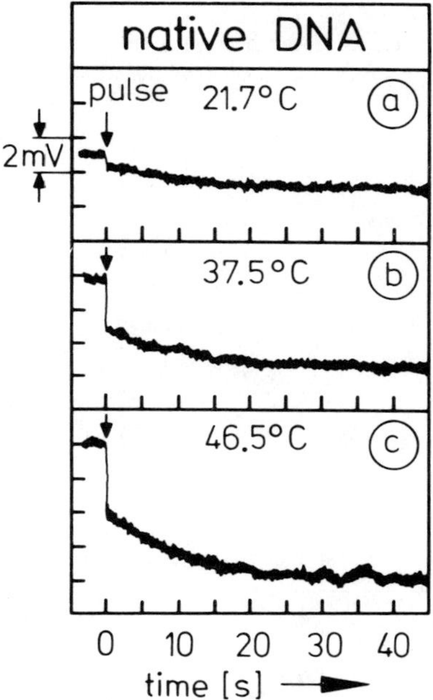

FIG. 20. Degradation of native DNA. Oscilloscope traces demonstrating the fast and slow modes of LSI decay at different temperatures. [DNA] = $7{\cdot}5 \times 10^{-2}$ g/litre in oxygen-saturated 10^{-2}M NaCl solution; absorbed dose per pulse = 50 krad.[16]

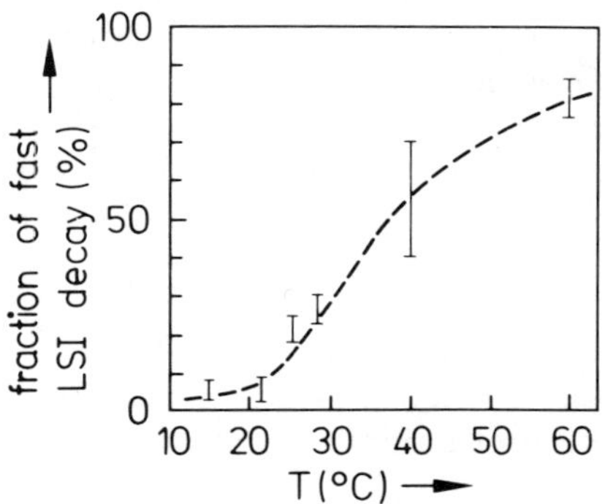

FIG. 21. Degradation of native DNA. Fraction of the fast mode of LSI decrease versus temperature. [DNA] = $7{\cdot}5 \times 10^{-2}$ g/litre in oxygen-saturated 10^{-2}M NaCl solution; absorbed dose per pulse = 50 krad.[16]

TABLE 3

DOUBLE STRAND SCISSION IN DNA. THE INFLUENCE OF TEMPERATURE ON THE RATE OF THE TWO MODES OF LSI DECREASE

Effects observed upon increasing the temperature	*Explanations*
Degree of degradation increases	h_{crit} Increases
$\tau_{1/2\,(slow)}$ becomes faster	Rate of melting of H-bonds increases
Fraction of fast LSI decrease is augmented	h_f Increases
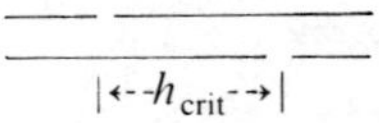 $\vert\leftarrow h_{crit} \rightarrow\vert$	If $h < h_{crit}$ double strand breaks occur
	If $h < h_f$ fast double strand breaks occur
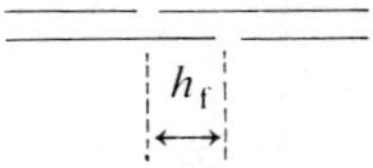 h_f $\leftrightarrow$	If $h_{crit} > h > h_f$ slow double strand breaks occur

degradation with increasing temperature is due to the fact that h_{crit} increases. As the rate of breaking of H-bonds increases $\tau_{1/2}(\text{LSI})_{slow}$ becomes faster. The increase in the fraction of the fast mode of LSI decrease is augmented since h_f increases. The increase in h_{crit} is also responsible for the occurrence of a third mode of LSI decay at temperatures above 40 °C. The third mode presumably corresponds to rather long fragments which detach relatively slowly, even at higher temperatures, but which do not detach at all at lower temperatures. The curve shown in Fig. 21 resembles the melting curves of double stranded oligonucleotides. Based on literature data (for example, see ref. 27) a midpoint of about 35 °C of the curve in Fig. 21 corresponds to a chain length of about 7–15 nucleotide units. This number may serve as a rough estimate for the length of detaching segments.

4.2.2. *Denatured DNA*

Denaturation of native DNA was performed by heating in solution with 10^{-2} M $NaClO_4$ for 10 min and subsequent rapid chilling. No minimum dose was necessary to detect a change of LSI in this case. The relative extent of LSI decrease $(U_0 - U_\infty)/(U_\infty - U_L)$ increased linearly with the absorbed dose per pulse. Only one mode of LSI decrease was observed if the native DNA was perfectly denatured (care had to be taken in order to prevent partial renaturation). $\tau_{1/2}(\text{LSI})$ was significantly smaller than for

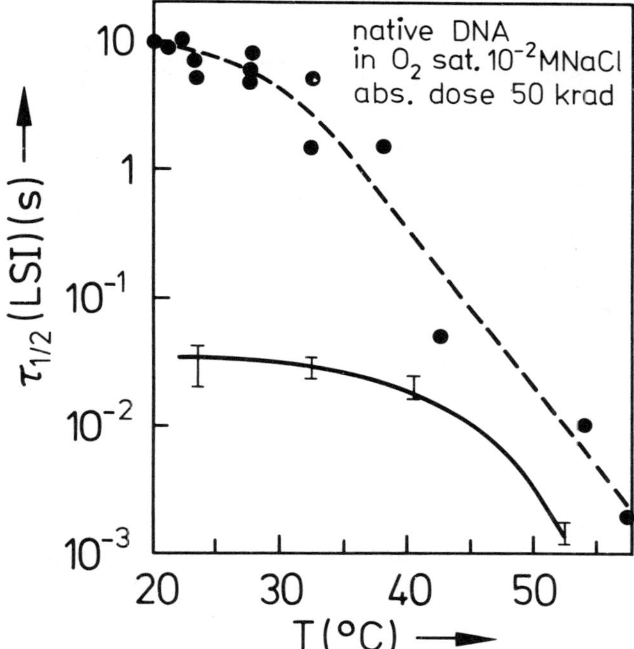

FIG. 22. Degradation of native DNA. Temperature dependence of $\tau_{1/2}$(LSI). Upper curve: 'slow' process; lower curve: 'fast' process. $\bar{M}_{w,0} = 1{\cdot}2 \times 10^7$; [DNA] = $7{\cdot}5 \times 10^{-2}$ g/litre in oxygen-saturated 10^{-2}M NaCl solution; absorbed dose per pulse = 50 krad.[16]

native DNA; it decreased continuously with increasing absorbed dose as shown in Fig. 23.

All results are consistent with the single stranded structure of the denatured polymer. The lowest absorbed doses applied during the experiments shown in Fig. 23 pertain to Z_s values of the order of 10. For technical reasons (signal-to-noise ratio) the measurements could not yet be extended to the range covering $0 < Z_s < 10$. Therefore, a comparison with polymers not bearing ionisable groups (dealt with in Section 4.1) is not yet possible. The decrease in $\tau_{1/2}$(LSI) with increasing dose is a consequence of the fact that smaller and therefore faster diffusing fragments are generated at higher doses. It appears that the decrease in $\tau_{1/2}$(LSI) with increasing absorbed dose is, in this case, much more drastic than in the case of PPVK. It therefore seems that due to their polyelectrolyte nature, single stranded

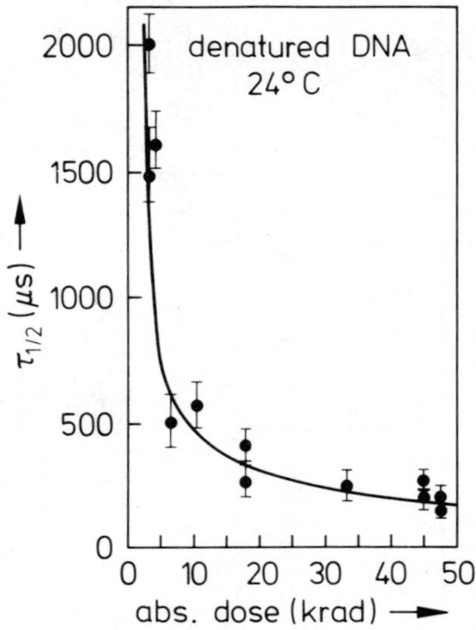

FIG. 23. Degradation of denatured DNA. $\tau_{1/2}$(LSI) as a function of the absorbed dose per pulse. Oxygen-saturated solution containing NaCl (10^{-2} mol/litre).[16]

DNA molecules are highly expanded. The separation of the fragments is not hindered significantly and may be considered as an essentially translatory diffusion process in contrast to fragment separation in the case of non-electrolyte polymers of coiled conformation.

5. STUDIES OF CHEMICAL REACTIONS

5.1. Macroradical Lifetimes and Reactions with Scavengers

5.1.1. *Pulse Radiolysis Studies on Poly(methylmethacrylate) (PMMA)*

Upon irradiating PMMA in acetone solution in absence of O_2, two modes of LSI decrease were observed as shown in Fig. 24.[17] From the results obtained it could be inferred that the two modes are owing to two different transient species. The rapid mode with a lifetime of about 20 μs was not influenced by additives. It was attributed to the diffusion of fragments. Main-chain rupture is in this case a very rapid process (much more rapid than fragment separation). It is assumed that the intermediate responsible for this process is a molecule in a repulsive (electronically) excited state or an ion. The slow mode of LSI decrease (lifetime: *ca.* 6 ms; ($k = 170\,s^{-1}$)) is

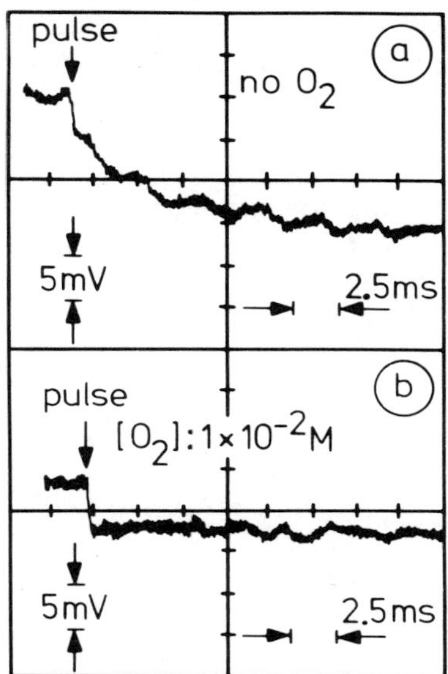

FIG. 24. Degradation of poly(methylmethacrylate) in acetone at 24 °C by 16 MeV electrons. Oscilloscope traces demonstrating the decrease in LSI after a 2 μs pulse. [PMMA] = $1{\cdot}4 \times 10^{-2}$ base mol/litre; absorbed dose per pulse = $4{\cdot}8 \times 10^4$ rad. (a) Ar-saturated solution, (b) $[O_2] = 1 \times 10^{-2}$ mol/litre.[17]

assumed to be related to the lifetime of chain side macroradicals. The extent of LSI decrease owing to the slow process and $\tau_{1/2}(\text{LSI})_{\text{slow}}$, decreased with increasing concentration of O_2 or C_2H_5SH added to acetone solutions. As shown in Fig. 24(b) the slow mode was completely suppressed at $[O_2] = 1 \times 10^{-2}$ mol/litre.

The kinetic treatment of the results was based on the mechanism presented in Table 4. It was assumed that in the case of the rapid process $k_7 \gg k_8$ and in the case of the slow process $k_9 \ll k_{10} \equiv k_8$. Reactions (11a) and (11b) inhibit main-chain rupture according to reaction (9). The degree of degradation, therefore, becomes smaller if the system contains scavenger molecules S. Pseudo-first-order kinetics can be applied at sufficiently high concentrations of S, that is, if $[S]$ is not significantly changed during the reaction.

The diminution of the concentration of chain side macroradicals P· as a function of time can be expressed in this case by eqn. (XVII)

$$[\text{P}\cdot]_t = [\text{P}\cdot]_0 \exp\left(-(k_9 + \Sigma k_{11}[S])t\right) \qquad \text{(XVII)}$$

By measuring $\tau(\mathrm{LSI})$ as a function of the scavenger concentration; k_{11} was obtained using eqn. (XVIII) (where $k_9 = 170\,\mathrm{s}^{-1}$):

$$\tau(\mathrm{LSI})^{-1} = k_9 + [S]\Sigma k_{11} \tag{XVIII}$$

The plot obtained from the experiments with oxygen is shown in Fig. 25. The following rate constants were obtained with acetone solutions for the reaction of O_2 and C_2H_5SH with the transient causing the slow main-chain

TABLE 4
MAIN-CHAIN DEGRADATION OF POLY(METHYLMETHACRYLATE) IN SOLUTION AND ITS INHIBITION BY SCAVENGERS

Elementary reaction	*Number*	*Depiction*
P^*_{n+m} (highly excited state or anion) $\xrightarrow{k_7}$ $(P^{\cdot}_n \ldots P^{\cdot}_m)$ (entangled fragments)	(7)	Fast main-chain scission
$(P^{\cdot}_n \ldots P^{\cdot}_m) \xrightarrow{k_8} P^{\cdot}_n + P^{\cdot}_m$	(8)	Diffusion of fragments
$P^{\cdot}_{n+m}$ (lateral macroradical) $\xrightarrow{k_9}$ $(P_n \ldots P^{\cdot}_m)$ (entangled fragments)	(9)	Slow main-chain scission
$(P_n \ldots P^{\cdot}_m) \xrightarrow{k_{10}} P_n + P^{\cdot}_m$	(10)	Diffusion of fragments
$P^{\cdot}_{n+m} + S \xrightarrow{k_{11}} P_{n+m}\text{—}S\cdot$	(11a)	
$P^{\cdot}_{n+m} + S \xrightarrow{k'_{11}} P_{n+m} + S\cdot$	(11b)	

scissions: $k_{O_2} = 1 \times 10^7$ litre/mol s and $k_{C_2H_5SH} = 2{\cdot}5 \times 10^4$ litre/mol s. The value of k_{O_2} is of the same magnitude as the value estimated from polymerisation studies for the reaction of chain end macroradicals of PMMA with O_2.[18]

5.1.2. *Pulse Radiolysis Studies on Poly(ethylene oxide) (PEO)*

The irradiation of aqueous PEO solutions in the absence of oxygen at a concentration of 1×10^{-2} base mol/litre caused crosslinking as indicated by the increase of the LSI after the pulse (Fig. 26). However, upon carrying out the irradiations at $[O_2] = 1{\cdot}4 \times 10^{-3}$ mol/litre under otherwise identical conditions, the LSI decreased after the pulse.[19] As shown in Fig. 27, this decrease consists of two modes. In the case of the rapid mode, plots

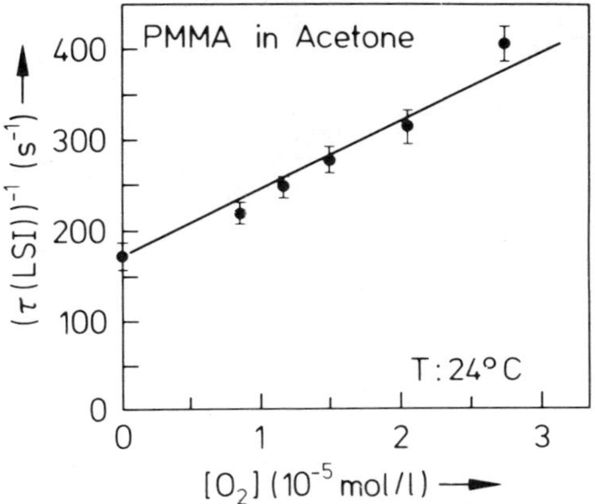

FIG. 25. Degradation of PMMA in acetone at 24 °C. Reciprocal lifetime of the slow LSI decrease versus the oxygen concentration. [PMMA] = $1{\cdot}4 \times 10^{-2}$ base mol/litre; absorbed dose per pulse $4{\cdot}8 \times 10^{4}$ rad.[17]

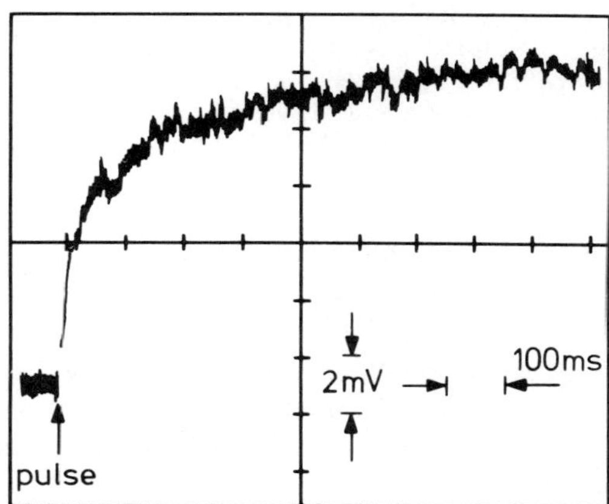

FIG. 26. Irradiation of poly(ethylene oxide) in an oxygen-free aqueous solution at room temperature. [PEO] = $5{\cdot}7 \times 10^{-2}$ base mol/litre; $\bar{M}_{w,0} = 4 \times 10^{6}$. Oscilloscope trace demonstrating the increase in LSI after irradiation with a 100 ns pulse of 16 MeV electrons. Absorbed dose per pulse = $1{\cdot}3 \times 10^{3}$ rad, $U_0 = 285$ mV.[19]

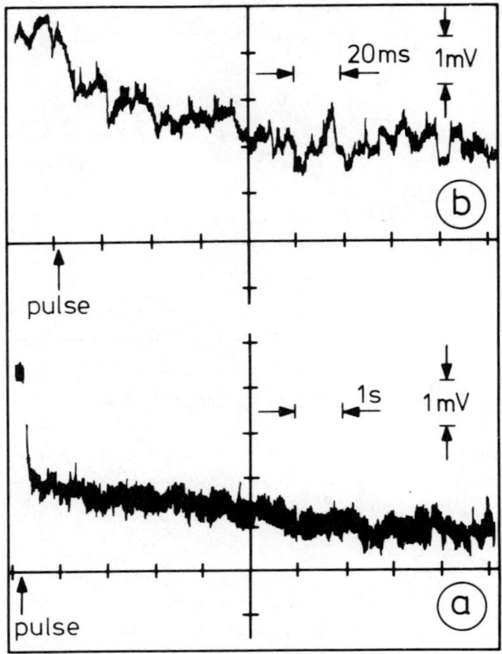

FIG. 27. Oxidative degradation of poly(ethylene oxide) in aqueous solution at room temperature. [PEO] = $5{\cdot}7 \times 10^{-2}$ base mol/litre; $[O_2] = 1{\cdot}4 \times 10^{-3}$ mol/litre. Oscilloscope traces demonstrating the decrease in LSI after irradiation with a 50 ns pulse of 16 MeV electrons. Absorbed dose per pulse = 600 rad; $U_0 = 239$ mV. (a) Total diminution of LSI; (b) time resolved rapid mode of LSI decrease.[19]

made according to eqn. (XII) yielded straight lines indicating that the rate of the LSI decrease is determined by a second order process. The first half-life time, $\tau_{1/2}(\text{LSI})_1$, decreased with increasing absorbed dose as shown in Fig. 28(a). A plot of the reciprocal first half-life time yielded a straight line (Fig. 28(b)). From these results it is inferred that the rate of the fast mode of LSI decrease is associated with a chemical process, in this case a bimolecular reaction.

These results can be explained in terms of the mechanism shown in Table 5. Macroradicals are generated upon hydrogen abstraction by OH radicals produced during the radiolysis of water. In the absence of O_2 the macroradicals combine (crosslinking). Reaction (13) competes with reaction (14) if O_2 is present in the solution. At sufficiently high

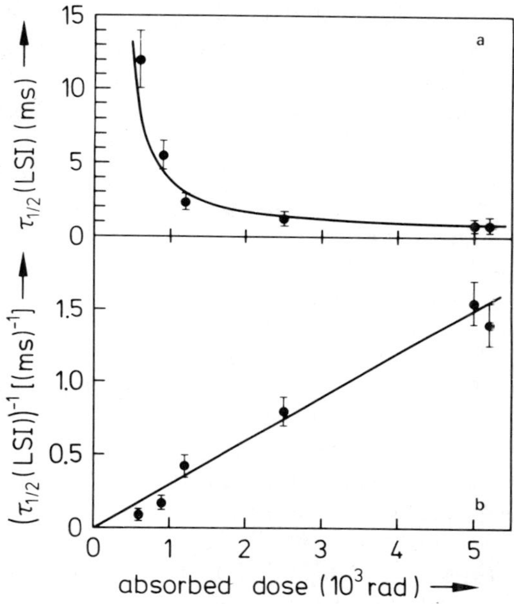

FIG. 28. Degradation of poly(ethylene oxide) in aqueous solution. $[O_2] = 1{\cdot}4 \times 10^{-3}$ mol/litre; $[\mathrm{PEO}] = 5{\cdot}7 \times 10^{-2}$ base mol/litre; $\bar{M}_{w,0} = 4 \times 10^6$. (a) $\tau_{1/2}$(LSI) and (b) $\tau_{1/2}(\mathrm{LSI})^{-1}$ versus the absorbed dose per pulse.[19]

TABLE 5

OXIDATIVE MAIN-CHAIN DEGRADATION OF POLY(ETHYLENE OXIDE) IN AQUEOUS SOLUTION

Elementary reaction	*Number*	*Depiction*
$PH + \cdot OH \rightarrow P\cdot + H_2O$	(12)	Abstraction of H-atoms
$P\cdot + P\cdot \rightarrow P{-}P$	(13)	Crosslinking
$P\cdot + O_2 \rightarrow PO_2^{\cdot}$	(14)	Formation of peroxyradicals
$PO_2^{\cdot} + PO_2^{\cdot} \rightarrow 2PO\cdot + O_2$	(15)	Formation of oxyradicals
$PO\cdot \rightarrow F_1 + F_2^{\cdot}$	(16)	Main-chain scission (decomposition of oxyradicals)

concentrations of O_2 reaction (13) is completely suppressed. Practically all macroradicals are converted to peroxyradicals according to reaction (14). During the combination of the latter, oxyradicals PO· are formed according to reaction (15). This reaction is considered as the rate determining step during the degradation process.

The subsequent decomposition of the oxyradicals should be comparably rapid. Presumably reaction (15) will involve a transient which, however, is also relatively shortlived.

The slow mode of LSI decrease (Fig. 27(a)) is assumed to be due to further main-chain scissions induced by macroradicals $F_2^{\cdot}$ generated during the decomposition of PO· radicals (reaction (16)). Hydrogen atoms can be abstracted by $F_2^{\cdot}$ from intact PEO molecules ($F_2^{\cdot} + PH \rightarrow F_2H + P$) or react with $O_2(F_2^{\cdot} + O_2 \rightarrow F_2O_2)$ followed by a reaction analogous to reaction (15). The occurrence of a chain reaction during the oxidative main-chain rupture has been inferred by Marchal and co-workers[20] from their results obtained by irradiating dilute aqueous PEO solutions with ^{60}CO-γ-rays. Evidence for the fact that the slow mode of LSI decrease is correlated with a chain reaction† is also derived from results concerning the dependence of the degree of degradation on the absorbed dose. Whilst the total degree of degradation increases, the fraction due to the slow process is steadily decreasing with increasing absorbed dose. Actually, this behaviour is expected. As the absorbed dose is increased the instantaneous radical concentration shortly after the pulse is increased. Therefore radical–radical reactions (chain termination reactions) will predominate over hydrogen abstraction reactions which propagate the chain reaction. The high energy induced oxidative degradation of PEO in acetonitrile solution proceeds in a similar way as in aqueous solution. In dilute, oxygen-free acetonitrile solutions of PEO, a main-chain rupture process was induced which involved a very fast chemical step. Therefore, the fragment diffusion was again determining the rate of LSI decrease after the pulse. In the latter case, $\tau_{1/2}$(LSI) amounted to about 10–20 μs and was only slightly dependent on the absorbed dose (at $Z_s \leq 1$). The extent of LSI decrease became greater with increasing dose according to eqn. (VII).

5.1.3. *Flash-photolysis Studies on Poly-α-methylstyrene* (*PαMS*) *and Polystyrene* (*PSt*)

On irradiating PαMS and PSt in chloroform or carbon tetrachloride solutions with 20 ns flashes of 265 nm light, LSI changes were observed

† The kinetic chain length is certainly very short in this case.

TABLE 6

DEGRADATION OF POLYSTYRENE IN AIR-SATURATED $CHCl_3$ SOLUTION AT ROOM TEMPERATURE. $\bar{M}_{w,0} = 2 \times 10^6$; $[PSt] = 1 \times 10^{-3}$ BASE MOL/LITRE. $\tau_{1/2}$(LSI) AT VARIOUS SCATTERING ANGLES ϑ: ABSORBED DOSE PER FLASH = 8×10^{-6} EINSTEIN/LITRE

ϑ	$\tau_{1/2}(LSI)$ (*ms*)
45°	0·57
75°	0·54
90°	0·56
105°	0·55
120°	0·56
135°	0·54

which were correlated with chemical reactions.[21,22] This assignment is based on the following findings:

(a) The values of $\tau_{1/2}$(LSI) are relatively long (in the ms range).
(b) Oxygen exerts a pronounced influence on $\tau_{1/2}$(LSI).
(c) Recent experiments[23] demonstrated that $\tau_{1/2}$(LSI) is independent of the scattering angle (Table 6). If the rate of LSI change is determined by fragment diffusion, $\tau_{1/2}$(LSI) is expected to decrease with increasing scattering angle ϑ, as in the case of PPVK. A typical oscilloscope trace obtained with PSt is presented in Fig. 29. In absence of O_2 no degradation was observed. On the other hand, PαMS underwent main-chain scission in $CHCl_3$ and CCl_4 solutions freed from O_2. In both cases first order plots according to eqn. (XI) yielded straight lines, as shown in Fig. 29 for PSt, only at relatively low O_2 concentrations. Curved plots were observed at $[O_2] > 2 \times 10^{-4}$ mol/litre. From these results it was inferred that in the O_2 concentration region under investigation a change was occurring in the process which determines the rate of LSI decrease.

A straightforward discussion of these results is complicated by the fact that the solvent plays an important role during main-chain rupture of both polymers. In *p*-dioxane and dichloromethane solutions no LSI change was detected. This is in accordance with the findings of Price and Fox[23,24] who reported ϕ(S) values for these solvents several orders of magnitude smaller than those for $CHCl_3$ and CCl_4 solutions. The very small changes of the LSI

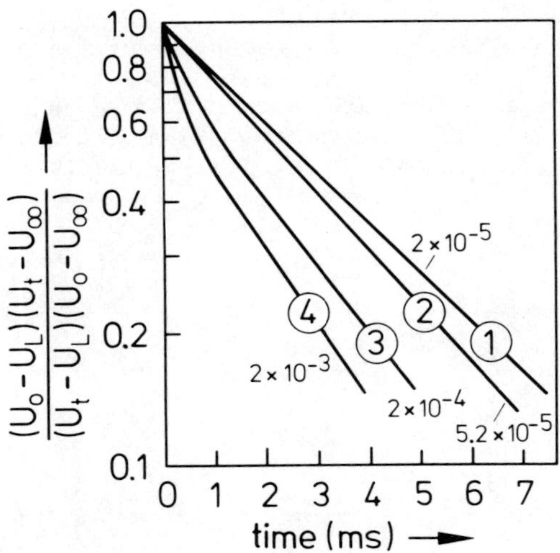

FIG. 29. Degradation of polystyrene in chloroform solution at room temperature. $\bar{M}_{w,0} = 2 \times 10^6$; [PSt] $= 9{\cdot}3 \times 10^{-4}$ base mol/litre. Dose absorbed by the solution per flash $= 3{\cdot}2 \times 10^{-6}$ Einstein/litre. First order plots according to eqn. (XI) for various oxygen concentrations (indicated on the curves in mol/litre).[22]

corresponding to those low ϕ(S) values could not be detected by our method.

A tentative reaction mechanism is presented in Table 7. It is assumed that macroradicals are primarily formed which can subsequently decompose by main-chain scission according to reaction (25), as in the case of PαMS, or react with oxygen in order to form an intermediate which fragments according to reaction (27) or (29) as in the case of PSt. (In reaction (27) $P^{\cdot}_{ox}$ represents a radical formed by a rearrangement process from $PO_2^{\cdot}$.)

At present it appears feasible that the solvent is involved in the formation of macroradicals via an exciplex of (possibly) charge-transfer character according to reactions (22) and (23) in Table 7. Processes involving solvent radicals according to reactions (18) and (19) presumably do not play a significant role (except for the initiation of chain reactions) since, due to the low value of the extinction coefficient of $CHCl_3$ at 265 nm ($\varepsilon_{265} = 8{\cdot}1 \times 10^{-4}$ litre/mol cm), only a very small portion of the incident light is absorbed by the solvent. The overwhelming portion is absorbed by the polymer (ε_{265}(PαMS) = 224 litre/(base mol) cm and ε_{265}(PSt) = 210 litre/(base mol) cm).

If one assumes that in the case of PαMS the rate of LSI decrease is

TABLE 7
PHOTOLYTIC MAIN-CHAIN DEGRADATION OF POLYSTYRENE AND POLY-α-METHYLSTYRENE IN CHLOROFORM SOLUTION (P AND S DESIGNATE POLYMER AND SOLVENT MOLECULES, RESPECTIVELY)

Elementary reaction	*Number*
Generation of macroradicals	
$P + h\nu \longrightarrow P^*$	(17)
$S + h\nu \longrightarrow S^*$	(18)
$S^* \longrightarrow 2R_S^{\cdot}$	(19)
$R_S^{\cdot} + P \longrightarrow R_SH + P\cdot$	(20)
$P^* + S \longrightarrow P + S^*$	(21)
$P^* + CHCl_3 \longrightarrow (P^* \ldots CHCl_3)$	(22)
$(P^* \ldots CHCl_3) \longrightarrow P\cdot + H\dot{C}Cl_2 + HCl$	(23a)
$(P^* \ldots CHCl_3) \longrightarrow P^{\cdot +} + (HCCl_3)^{\cdot -}$; $P^{\cdot +} \downarrow P\cdot + H^+$	(23b)
Reactions of macroradicals	
$P\cdot + O_2 \longrightarrow PO_2^{\cdot}$	(24)
$P\cdot \longrightarrow F_1 + F_2^{\cdot}$	(25)
$PO_2^{\cdot} \longrightarrow P_{ox}^{\cdot}$	(26)
$P_{ox}^{\cdot} \longrightarrow F_1 + F_2$	(27)
$2PO_2^{\cdot} \longrightarrow 2PO\cdot + O_2$	(28)
$PO\cdot \longrightarrow F_1 + F_2^{\cdot}$	(29)
$P\cdot + P\cdot \longrightarrow$ products	(30)
$P\cdot + R_S^{\cdot} \longrightarrow$ products	(31)

associated with reaction (25) and that at low O_2 concentration reactions (24) and (25) are competing, the following rate constants are obtained: $k_{25} = 3{\cdot}5 \times 10^2\ s^{-1}$ and $k_{24} = 5{\cdot}5 \times 10^5$ litre/mol s. For PSt $k_{24} = 5 \times 10^5$ litre/mol s was estimated and the composite rate constant of the decay of macroradicals P· in the absence of oxygen was $2{\cdot}4 \times 10^2\ s^{-1}$. The values of k_{24} appear to be appreciably lower than expected on the basis of encounter-controlled reactions. It therefore follows that in this case additional information has to be gained for an unambiguous correlation of the rate of LSI change with a specific chemical reaction.

This section may be concluded by pointing out that the function LSI = f(t) could be correlated in certain cases with the rate of chemical reactions associated with main-chain degradation processes. As a consequence of the complexity of most degradation processes the correlation of τ(LSI) is not a simple problem as long as the mechanism of the degradation process has not been elucidated. In the examples dealt with in this chapter (*vide ante*) we have tried to derive from measured rates of LSI decrease, mechanistic as well as kinetic conclusions. In general, the light scattering detection method can be a powerful tool for gaining kinetic data on chemical reactions associated with main-chain degradation processes, if its application is supported by other methods capable of elucidating the mechanism of degradation.

5.2. 'Simultaneous' Degradation and Crosslinking

At the beginning of this chapter the versatility of the light scattering detection method was pointed out and the reader might subsequently have obtained some insight into the capability of the method in the investigation of quite different kinds of problems. The reader's attention may now be drawn to a rather curious aspect which concerns the high time-resolution of LSI measurements now available. This aspect deals with the phenomenon of 'simultaneous' crosslinking and degradation which is encountered frequently if a polymer is exposed to an external agency, such as high energy or u.v. light irradiation, mechanical stress, heat, etc. under steady-state conditions.

In a study concerning the high energy irradiation of poly(methylvinylketone) in solution, it was found that the polymer undergoes crosslinking and main-chain scission. Depending on the polymer concentration one or other of the two processes predominates (an analogous behaviour has been found with many other polymers).[26] If poly(methylvinylketone) was irradiated with 2 μs pulses of 16 MeV electrons,[25] the LSI decreased immediately after the pulse with $\tau_{1/2}(\mathrm{LSI})_{\mathrm{decr.}} \approx 20\ \mu\mathrm{s}$. Subsequently an increase was observed ($\tau_{1/2}(\mathrm{LSI})_{\mathrm{inc.}} \approx 0{\cdot}4\ \mathrm{s}$). A typical oscilloscope trace is presented in Fig. 30, which shows that the processes of main-chain scission (immediately after the pulse) and crosslinking (at a later stage) could be discriminated in this case. The rate of the LSI decrease pertaining to the main-chain scission process was proved to be determined by disentanglement diffusion, whereas the rate of the LSI increase was correlated with the rate of a chemical reaction. A similar situation was recently encountered in

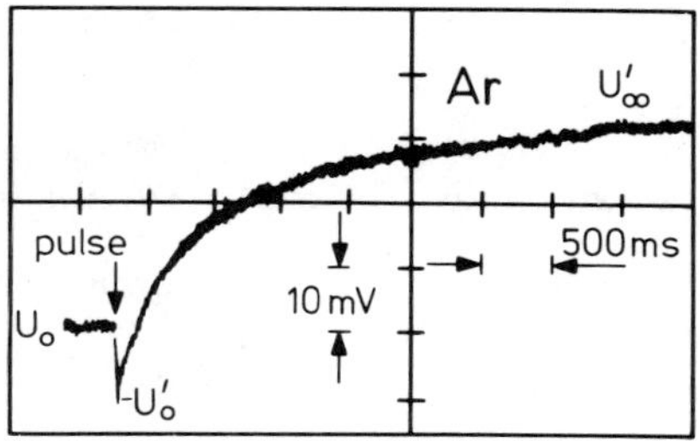

FIG. 30. LSI changes observed after irradiating an Ar-saturated acetone solution of poly(methylvinylketone) with a 2 μs pulse of 16 MeV electrons at room temperature. $\bar{M}_{w,0} = 3{\cdot}9 \times 10^6$; [PMVK] $= 2{\cdot}86 \times 10^{-2}$ base mol/litre; absorbed dose per pulse $= 2 \times 10^4$ rad.[25]

the case of poly(ethylene oxide) dissolved in aqueous solutions of fairly low oxygen concentration.[19] These examples show that time-resolved LSI measurements can be rather helpful in obtaining mechanistic information which is otherwise scarcely available or unavailable.

ACKNOWLEDGEMENT

The author of this chapter wishes to express his acknowledgement to Dr G. Beck who co-operated enthusiastically and patiently as an electronics specialist with polymer chemists and contributed greatly to the progress of these investigations. The speed of progress is always critically dependent on the availability of appropriate compounds. In our case Mrs U. Fehrmann synthesised many polymer samples which is gratefully acknowledged here. Last, but not least, our extremely skilful secretary, Mrs G. Snoei, is to be mentioned. Without her diligence and accuracy in typing the various manuscripts nobody would ever have learned about our work.

REFERENCES

1. BECK, G., KIWI, J., LINDENAU, D. and SCHNABEL, W., *Eur. Polym. J.*, **10** (1974) p. 1069.
2. BECK, G., LINDENAU, D. and SCHNABEL, W., *Eur. Polym. J.*, **11** (1975) p. 761.
3. BECK, G., DOBROWOLSKI, G., KIWI, J. and SCHNABEL, W., *Macromolecules*, **8** (1975) p. 9.
4. LINDENAU, D. and SCHNABEL, W., *J. Polym. Sci., Polym. Lett. Ed.*, **16** (1978) p. 665.

5. TAGAWA, S. and SCHNABEL, W., *Makromol. Chem.*, in press.
6. BROCHARD, E. and DEGENNES, P. G., *Macromolecules*, **10** (1977) p. 1157.
7. KUNST, E. D., Rec. Trav. Chim., *Pays-Bas Belg.*, **69** (1950) p. 125.
8. TAGAWA, S. and SCHNABEL, W., *Macromolecules*, in press.
9. DOBROWOLSKI, G., KIWI, J. and SCHNABEL, W., *Eur. Polym. J.*, **12** (1976) p. 657.
10. DOBROWOLSKI, G. and SCHNABEL, W., *Z. Naturforsch.*, **31a** (1976) p. 109.
11. DOBROWOLSKI, G. and SCHNABEL, W., unpublished results.
12. CHAUFER, B., SEBILLE, B. and QUIVORON, C., *Eur. Polym. J.*, **11** (1975) pp. 683, 689, 695.
13. DONDOS, A., *C. R. Acad. Sc.*, Paris, Ser. C, **272** (1971) p. 1419.
14. DONDOS, A. and BENOIT, H., *J. Polym. Sci., Polym. Phys. Ed.*, **15** (1977) p. 137.
15. LINDENAU, D., HAGEN, U. and SCHNABEL, W., *Z. Naturforsch.*, **31c** (1976) p. 484.
16. LINDENAU, D., HAGEN, U. and SCHNABEL, W., *Rad. Environm. Biophys.*, **13** (1976) p. 287.
17. BECK, G., LINDENAU, D. and SCHNABEL, W., *Macromolecules*, **10** (1977) p. 135.
18. SCHULZ, G. V. and HENRICI, G., *Makromol. Chem.*, **18/19** (1956) p. 437.
19. GRÖLLMANN, U. and SCHNABEL, W., in preparation.
20. (a) CROUZET, C. and MARCHAL, J., *Makromol. Chem.*, **166** (1973) pp. 69, 85, 99.
    (b) DECKER, C. and MARCHAL, J., *Makromol. Chem.*, **166** (1973) pp. 117, 139, 155.
21. BEAVAN, S. W., BECK, G. and SCHNABEL, W., *Eur. Polym. J.*, **14** (1978) p. 385.
22. BEAVAN, S. W. and SCHNABEL, W., *Macromolecules*, **11** (1978) p. 782.
23. PRICE, T. R. and FOX, R. B., *J. Polym. Sci., Polym. Lett. Ed.*, **4** (1966) p. 771.
24. FOX, R. B. and PRICE, T. R., *J. Polym. Sci.*, **A3** (1965) p. 2303.
25. LINDENAU, D., BEAVAN, S. W., BECK, G. and SCHNABEL, W., *Eur. Polym. J.*, **13** (1977) p. 819.
26. SCHNABEL, W. In *Aspects of Degradation and Stabilization of Polymers*, H. H. G. Jellinek, Ed., Amsterdam, Elsevier, 1978, p. 149.
27. PÖRSCHKE, D. and EIGEN, M., *J. Mol. Biol.*, **62** (1971) p. 361.

Chapter 3

THERMAL DEGRADATION OF POLYESTERS IN THE MASS SPECTROMETER

INGO LÜDERWALD

University of Mainz, Mainz, West Germany

SUMMARY

Mass spectrometry is used to investigate the thermal degradation reactions of polyesters. The polymer samples are pyrolysed directly in the ion source of the instrument and their volatile pyrolysis products detected immediately after ionisation. The high sensitivity of a mass spectrometer allows low degradation rates ($\leq 1\%$/min) and low pyrolysis temperatures at which only the activation energies of selective pyrolysis processes are achieved.

Polyesters of aliphatic and aromatic dicarboxylic acids and different diols are preferentially degraded by thermal cis-*elimination or cleavage of the ester bond yielding pyrolysis products with carboxylic and alkene or ketene and hydroxyl end groups.*

Polylactones show a different degradation behaviour dependent on the ring size of the corresponding lactones (monomers) and their substituents. Thermal cis-*elimination, cleavage of the ester bond as well as the degradation via cyclic oligomers is found.*

INTRODUCTION

Thermal degradation processes of polymers have been extensively investigated during the last 40 years[1-3] and as early as 1860 a pyrolysis experiment was reported relating to the preparation of isoprene from natural rubber.[4] As well as the pure analytical application of pyrolysis, which aims at elucidating the primary structure of a polymer or copolymer, numerous studies have been devoted to the chemistry of thermal degradation processes and their mechanisms.

A classical procedure in this area of research has been the preparative pyrolysis of polymers with the trapping of volatile pyrolysis products at different temperatures, e.g. at −80 °C and −190 °C, followed by further fractionation and identification of the separated products by classical analytical methods (formation of derivatives, determination of melting points) or physical methods of instrumental analysis (IR, NMR, GC, MS). It is not the purpose of this paper to discuss the advantages and disadvantages of these pyrolysis procedures. Instead, I will attempt to summarise the information on the thermal degradation of various polyesters which can be obtained by using the technique of direct pyrolysis in the ion source of a commercial mass spectrometer.[5]

PYROLYSIS OF POLYMERS IN A MASS SPECTROMETER

In a mass spectrometer with electron impact as ionisation mode, organic compounds are volatised into the gas phase and ionised by interaction with an electron beam. Since the energy transmitted to the molecules normally far exceeds their ionisation potential (9–13 eV),[6] the molecular ions are further fragmented. However, these fragmentation reactions depend strongly upon the functional groups present. They are characteristic for various classes of organic compounds and often allow their identification.

If polymers are brought into the ion source of a mass spectrometer, they can be neither evaporated nor ionised. Upon heating a polymeric sample, the temperature of the initial thermal decomposition can be determined very accurately, since any volatile pyrolysis products that are produced will be ionised and detected as total ion current (TIC) before leaving the ion source. As the sensitivity of a mass spectrometer is in the region of 10^{-9} g, a sufficient amount of pyrolysis products are evaporated at degradation rates of about 1 %/min (1 %/min $\simeq 10^{-7}$ g, starting from 0·1 mg polymer sample).

Such low degradation rates obtained at low pyrolysis temperatures result in more selective pyrolysis mechanisms, which increases the possibility of detecting characteristic structural elements in the pyrolysis products. These low rates imply that only a few bonds are cleaved and this increases the probability that the pyrolysis products still contain the basic polymer structure, one, two or several units in length (eqn. 1).

$$\text{polymer} \xrightarrow{\Delta} \text{R} + \text{oligomers} + \text{R} \xrightarrow{Z_e} \begin{cases} \text{'molecular ions'} \\ \text{fragments} \end{cases} \quad (1)$$

The subsequent electron-impact ionisation yields positive ions of the pyrolysis products (molecular ions) as well as their fragment ions. Although the electron-impact fragmentation complicates the resulting mass spectral pattern, many primary fragmentation mechanisms of low molecular weight organic compounds (e.g. esters) are well known from the literature, so that at least one of the two initially unknown, thermally formed end groups can frequently be identified from these data (eqn. 1). If the structure of the end groups R and R′ can be elucidated from the resulting mass spectra, information about the pyrolysis mechanism becomes available. In addition, the high vacuum in the mass spectrometer ($P < 10^{-6}$ mm Hg) reduces the possibility of secondary thermal degradation reactions occurring because the pyrolysis products are immediately removed from the hot region and are ionised and detected within less than one second.

This model of direct pyrolysis in a mass spectrometer (eqn. 1) implies that a polyester like poly(ethyleneterephthalate) should be thermally degraded into a mixture of oligomeric esters (eqn. 2).

$$-(\text{CO—Ph—CO—O—CH}_2\text{—CH}_2\text{—O})- \xrightarrow{\Delta}$$

$$\text{R}-(\text{CO—Ph—CO—O—CH}_2\text{—CH}_2\text{—O})_{1,2,3\ldots}\text{R}' \quad (2)$$

Low molecular weight esters of terephthalic acid are known to undergo an α-cleavage into carboxonium ions as a characteristic fragmentation reaction[7,8] followed by a partial elimination of carbon monoxide (eqn. 3).

$$\text{R—O—CO—Ph—C}(=\text{O}^{\cdot+})\text{—O—R} \xrightarrow[-\cdot\text{OR}]{} \text{R—O—CO—Ph—C}\equiv\text{O|}^{+} \xrightarrow{-\text{CO}} \text{R—O—CO—Ph}^{+} \quad (3)$$

Therefore, if a mixture of thermally formed terephthalic esters (eqn. 2) is further fragmented into the ions shown in eqn. 3, their masses allow the identification of the structure of the end group R and the corresponding pyrolysis mechanism.

THERMAL DEGRADATION OF POLY(ALKYLENETEREPHTHALATES)

The thermal degradation mechanisms of poly(ethyleneterephthalate) (**Ia**), poly(d_4-ethyleneterephthalate) (**Ib**), poly(trimethyleneterephthalate) (**Ic**) and poly(tetramethyleneterephthalate) (**Id**) have been investigated in a mass spectrometer.[9,10] The pyrolysis-mass spectrum of **Ia**, obtained at 350 °C, is shown in Fig. 1.

The molecular weight of the structural unit (MW_{unit}) is 192, and the differences of 192 mass units between the peaks at *m/e* 917, 725, 533, 341 and 149 indicate that a mixture of oligomeric ions is formed which originate by the same pyrolysis and fragmentation mechanisms. The fragment *m/e* 149 is assigned as the carboxonium ion of terephthalic acid. Since in previous studies terephthalic acid was identified as one pyrolysis product of **Ia**,[11,12] it can be deduced that the series *m/e* 149–917 corresponds to oligomers with a thermally formed carboxyl end group (eqn. 4).

$$\mathrm{H{-}(O{-}CO{-}Ph{-}CO{-}O{-}CH_2{-}CH_2)_x{-}O{-}CO{-}Ph{-}C{\equiv}O|^+} \quad (4)$$

$m/e(x)$: *149*(0); *341*(1); *533*(2); *725*(3); *917*(4).

(i.e. when x equals 0, 1, 2, 3 and 4, *m/e* equals 149, 341, 533, 725 and 917, respectively.)

A likely pyrolysis mechanism is the thermal *cis*-elimination (eqn. 5) which must simultaneously yield carboxonium ions with a thermally formed vinylester end group (eqn. 6).

$$\mathrm{{\sim}C({=}O){-}C_6H_4{-}C({=}O){-}O{-}CH(H){-}CH_2{-}O{-}C({=}O){-}C_6H_4{-}C({=}O){\sim}} \quad (5)$$

$$\xrightarrow{\Delta}$$

$$\mathrm{{\sim}C({=}O){-}C_6H_4{-}C({=}O){-}O{-}CH{=}CH_2 + HO{-}C({=}O){-}C_6H_4{-}C({=}O){\sim}} \quad (6)$$

$-e$ (right product) → eqn. 4

$-e$ (left product) ↓

$$\mathrm{^+|O{\equiv}C{-}Ph{-}CO{-}(O{-}CH_2{-}CH_2{-}O{-}CO{-}Ph{-}CO)_x{-}O{-}CH{=}CH_2}$$

$m/e(x)$: *175*(0); *367*(1); *559*(2).

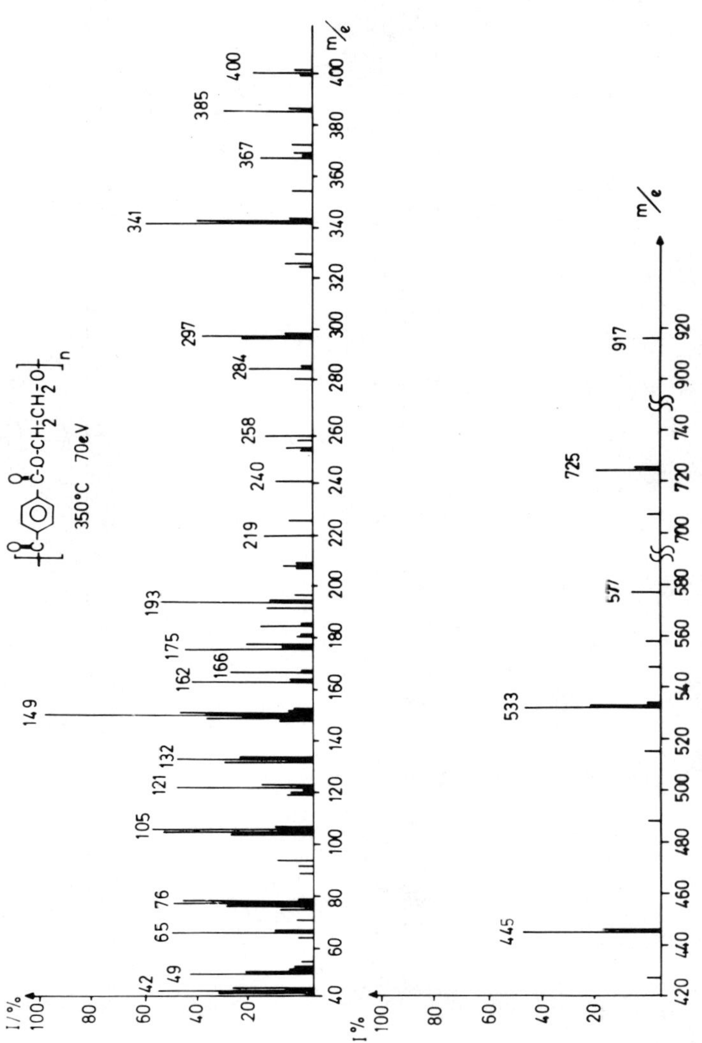

FIG. 1. Pyrolysis-mass spectrum of poly(ethyleneterephthalate) (**Ia**) at 350 °C.

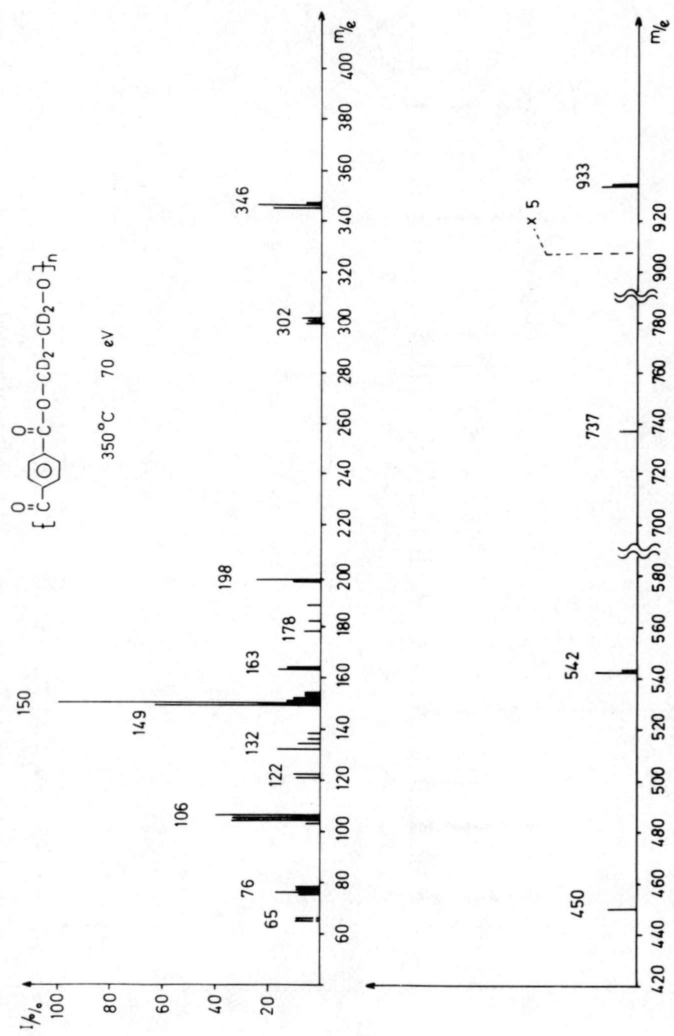

FIG. 2. Pyrolysis-mass spectrum of poly(d_4-ethyleneterephthalate) (**Ib**) at 350 °C.

In fact, these fragments are found in the pyrolysis-mass spectrum of **Ia** (Fig. 1). Furthermore, the proposed *cis*-elimination can be confirmed by an investigation of the deuterated polyester **Ib** (Fig. 2).

The expected shift of $(4x + 1)$ mass units confirms the structure of the fragments shown in eqn. 4 with x intact structural units (eqn. 7).

$$\text{PO}\!\left(\text{CO—Ph—CO—O—CR}_2\text{—CR}_2\text{—O}\right)_x\text{CO—Ph—C}{\equiv}\text{O|}^+ \quad (7)$$

R = H	$m/e(x)$:	*149*(0);	*341*(1);	*533*(2);	*725*(3);	*917*(4).
R = D		*150*(0);	*346*(1);	*542*(2);	*738*(3);	*934*(4).

Aromatic acids are known to eliminate carbon dioxide at high temperatures. For the polyesters under discussion it can be confirmed that the carboxyl end groups formed at 350 °C (eqn. 5) are also partially decarboxylated since this would explain the occurrence of the ions m/e 105, 297 and 489 (eqns. 8, 9).

$$\begin{array}{l}\text{H} \quad \text{Ph—CO—O—CH}_2\text{—CH}_2\text{—O—CO—Ph—CO—}\cdots \\ | \qquad\ \ | \\ \text{O—CO}\end{array} \quad (8)$$

$$\downarrow \Delta$$

$$\text{H—Ph—CO—O—CH}_2\text{—CH}_2\text{—O—CO—Ph—CO—}\cdots$$

$$\downarrow -e \qquad (9)$$

$$\text{H—Ph}\!\left(\text{CO—O—CH}_2\text{—CH}_2\text{—O—CO—Ph}\right)_x\text{C}{\equiv}\text{O|}^+$$

$m/e(x)$: *105*(0); *297*(1); *489*(2).

A second predominant pyrolysis reaction of poly(ethyleneterephthalate) is the cleavage of the ester bond with the formation of hydroxyl end groups (eqn. 10). The concomitant migration of a hydrogen atom from the adjacent glycol unit could be established by comparison with the deuterated polyester **Ib**.

$$\sim\text{CH}_2\text{—CH}_2\text{—O—}\overset{\text{O}}{\overset{\|}{\text{C}}}\text{—C}_6\text{H}_4\text{—}\overset{\text{O}}{\overset{\|}{\text{C}}}\,\wr\,\text{O—CH}_2\text{—CH}_2\text{—O}\sim \quad (\sim\text{H}\cdot \text{ migration}) \qquad (10)$$

$$\downarrow \Delta$$

$$\text{HO—CH}_2\text{—CH}_2\text{—O}\sim$$

Corresponding ions with a carboxonium end group formed by fragmentation according to eqn. 3 arise at *m/e* 193, 385 and 577 (Fig. 1; eqn. 11).

$$\mathrm{H{-}\!\left(O{-}CH_2{-}CH_2{-}O{-}CO{-}Ph{-}CO\right)_{\!x}\!{-}O{-}CH_2{-}CH_2{-}O{-}CO{-}Ph{-}C{\equiv}O|^+} \tag{11}$$

$$m/e(x):\quad \mathit{193}(0);\quad \mathit{385}(1),\quad \mathit{577}(2).$$

Furthermore, the small peaks at 445, 450, 459 and 473 found in the mass spectra of **Ia**, **Ib**, **Ic** and **Id**[9] indicate the thermal elimination of acetaldehyde[11,12] from poly(ethyleneterephthalate), which has been proposed to occur via the reaction of a carboxyl and a vinylester end group (eqns. 12 and 13).

$$\cdots\mathrm{{-}CO{-}Ph{-}COOH + CH_2{=}CH{-}O{-}CO{-}Ph{-}}\cdots \xrightarrow[-\mathrm{CH_3{-}CHO}]{\Delta}$$

$$\cdots\mathrm{{-}CO{-}Ph{-}CO{-}O{-}CO{-}Ph{-}}\cdots \tag{12}$$

$$\xrightarrow{-e^-} \mathrm{Ph{-}CO{-}O{-}CO{-}Ph{-}CO{-}O{-}R{-}O{-}CO{-}Ph{-}C{\equiv}O|^+} \tag{13}$$

$$\mathrm{R} = -\!\left(\mathrm{CH_2}\right)_2\!-,\quad m/e = 445;\qquad \mathrm{R} = -\!\left(\mathrm{CD_2}\right)_2\!-;\quad m/e = 450;$$

$$\mathrm{R} = -\!\left(\mathrm{CH_2}\right)_3\!-,\quad m/e = 459;\qquad \mathrm{R} = -\!\left(\mathrm{CH_2}\right)_4\!-;\quad m/e = 473.$$

Poly(trimethyleneterephthalate) (**Ic**) and poly(tetramethyleneterephthalate (**Id**) show corresponding thermal degradation behaviour at 340 °C with *cis*-elimination and cleavage of the ester bond as the predominant pyrolysis mechanisms.

POLYESTERS OF ALIPHATIC DICARBOXYLIC ACIDS AND DIOLS

Polyesters of adipic and succinic acid and various aliphatic diols generally exhibit a lower thermal stability than aromatic polyesters.[13] In Fig. 3 the pyrolysis-mass spectra of poly(ethyleneadipate) (**IIa**) and poly(d_4-ethyleneadipate) (**IIb**), measured at 280 °C, are compared.

The molecular weight of the structural unit of polyester **IIa** is 172, and the series of peaks at *m/e* 173, 345 (*m/e* 173 + 172) and 517 (*m/e*

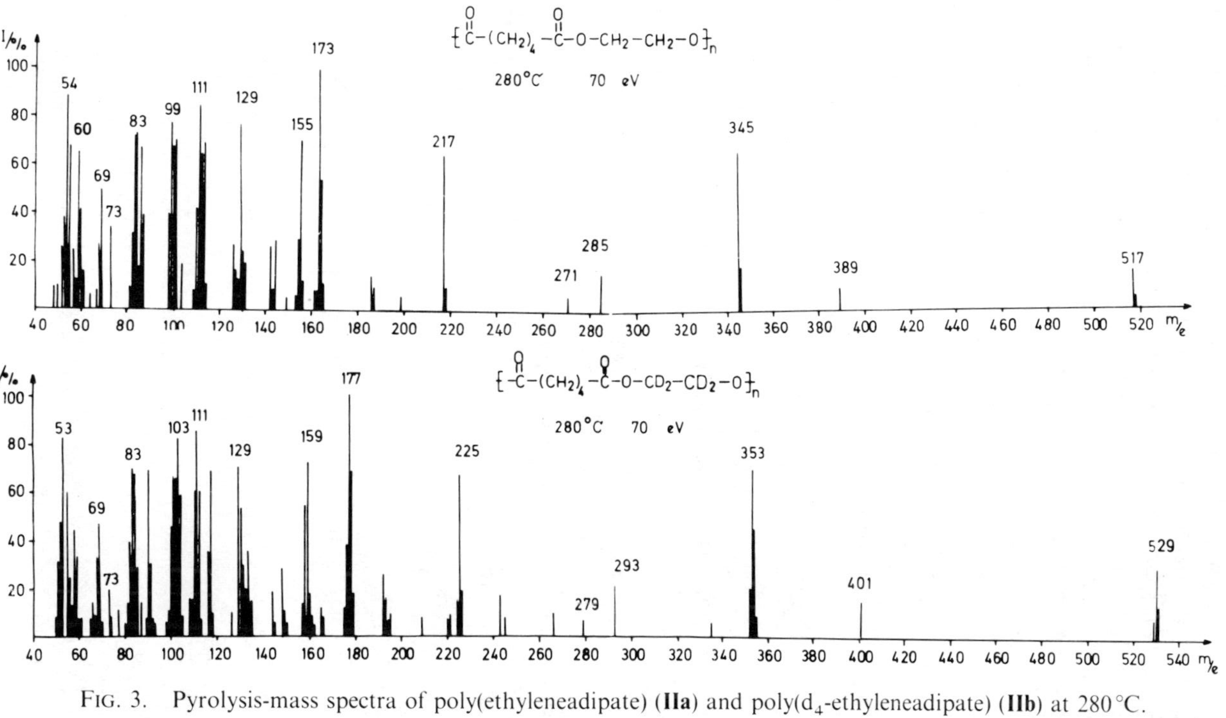

FIG. 3. Pyrolysis-mass spectra of poly(ethyleneadipate) (**IIa**) and poly(d_4-ethyleneadipate) (**IIb**) at 280 °C.

173 + (2 × 172)) indicate the presence of oligomeric fragments. The mass spectrum of the deuterated polyester **IIb** shows a shift of these ions to *m/e* 177 (+4), 353 (+8) and 529 (+12), which confirms that they contain 1, 2 and 3 d_4-ethylenediol units. The formation of these fragments can be explained by thermal cleavage of the ester bond (eqn. 14) and subsequent fragmentation into carboxonium ions analogous to those discussed above (eqn. 15).

$$\cdots\text{—CO—(CH}_2)_3\text{—CH(H)—C—O—CR}_2\text{—CR}_2\text{—O—CO—(CH}_2)_4\text{—CO—}\cdots \quad (14)$$

$$\downarrow \Delta$$

$$\cdots\text{—CO—(CH}_2)_3\text{CH=C=O + HO—CR}_2\text{—CR}_2\text{—O—CO—(CH}_2)_4\text{—CO—}\cdots$$

$$\downarrow -e \quad (15)$$

$$\text{H}\!\left(\text{O—CR}_2\text{—CR}_2\text{—O—CO}\!\left(\text{CH}_2\right)_4\text{CO}\right)_x\text{O—CR}_2\text{—CR}_2\text{—O—CO}\!\left(\text{CH}_2\right)_4\text{C}\equiv\text{O}|^+$$

R = H	*m/e*(*x*):	*173*(0);	*345*(1);	*517*(2).
R = D		*177*(0);	*353*(1);	*530*(2).

A further fragmentation process involving cleavage of alkyl—O bonds must explain the occurrence of ions at *m/e* 217 and 389 (**IIa**) and 225 and 401 (**IIb**) (eqn. 16).

$$\cdots\text{—CH}_2\text{—C(=O}^{+\cdot}\text{)—O—CH(H)—CH}_2\text{—O—}\cdots \rightarrow \cdots\text{—CH}_2\text{—C(=}^+\text{OH)—O—CH=CH}_2 \quad (16)$$

$$\downarrow$$

$$\text{H}\!\left(\text{O—CR}_2\text{—CR}_2\text{—O—CO}\!\left(\text{CH}_2\right)_4\text{CO}\right)_x\text{O—CR}_2\text{—CR}_2\text{—O—CO}\!\left(\text{CH}_2\right)_4\text{C(=}^+\text{OH)—O—CH=CH}_2$$

R = H	*m/e*(*x*):	*217*(0);	*389*(1).
R = D		*225*(0);	*401*(1).

Relatively low intensity peaks due to the ions *m/e* 129 and 155 (**IIb**: *m/e* 130 and 158) result from a thermal *cis*-elimination reaction (eqn. 17).

$$\cdots\text{—CO—(CH}_2\text{)}_4\text{—C(=O)—O—CR}_2\text{—CR(R)—O—CO—(CH}_2\text{)}_4\text{—CO—}\cdots$$

$$\downarrow\Delta \qquad\qquad \downarrow\Delta \qquad (17)$$

$$\cdots\text{—CO—(CH}_2\text{)}_4\text{—CO—OR} \qquad \text{CR}_2\text{=CR—O—CO—(CH}_2\text{)}_4\text{—CO—}\cdots$$

$$\downarrow -e \qquad\qquad \downarrow -e$$

$$^{+}|\text{O}\equiv\text{C—(CH}_2\text{)}_4\text{—CO—OR} \qquad \text{CR}_2\text{=CR—O—CO—(CH}_2\text{)}_4\text{—C}\equiv\text{O}|^{+}$$

m/e(R): *129*(H); *130*(D). *m/e*(R): *155*(H); *158*(D).

The peaks at *m/e* 111, 155 and 159 (**IIb**) in Fig. 3 are due to fragments with a thermally formed ketene end group (cf. eqn. 14) as shown in eqn. 18.

$$\text{O=C=CH—(CH}_2\text{)}_3\text{—C(=}^{+}\text{OH)—O—CR=CR}_2 \qquad \text{O=C=CH—(CH}_2\text{)}_3\text{C}\equiv\text{O}|^{+}$$

m/e(R): *155*(H); *159*(D). *m/e* 111.

(18)

The pyrolysis-mass spectra of poly(trimethyleneadipate), poly(tetramethyleneadipate), poly(ethylenesuccinate), poly(tetramethylenesuccinate) and poly(2,2-dimethyl-trimethylenesuccinate) reveal a similar thermal degradation behaviour.[13] In all cases the thermal cleavage of the ester bond into ketene and hydroxyl end groups is the prevailing pyrolysis process, and is accompanied by a less intensive *cis*-elimination. In addition, for polysuccinates a cleavage within dicarboxylic acid units becomes more significant, which may be due to the possibility for resonance stabilisation of the resulting acryloyl end groups (eqn. 19).

$$\cdots\text{—O—R—O—CO—CH}_2\text{—CH}_2\text{—CO—}\cdots \xrightarrow{\Delta}$$
$$\cdots\text{—O—R—OH + CO + CH}_2\text{=CH—CO—}\cdots \qquad (19)$$

THERMAL DEGRADATION OF POLYLACTONES

Polylactones are thermally even less stable than aliphatic polyesters of dicarboxylic acids and diols. Depending upon the ring size of the corresponding monomeric lactones poly-α-,† -β-, -δ- and -ε-lactones show typical differences in their degradation reactions.

Although poly-α-lactones are known to depolymerise on heating into six-membered ring cyclic dimers, the pyrolysis-mass spectra of poly(1,1-oxycarbonylethylene) (**III**) (Fig. 4) and poly(oxycarbonylmethylene)[14] exhibit, at 240 °C and P < 10^{-6} mm Hg, peaks up to *m/e* 700.

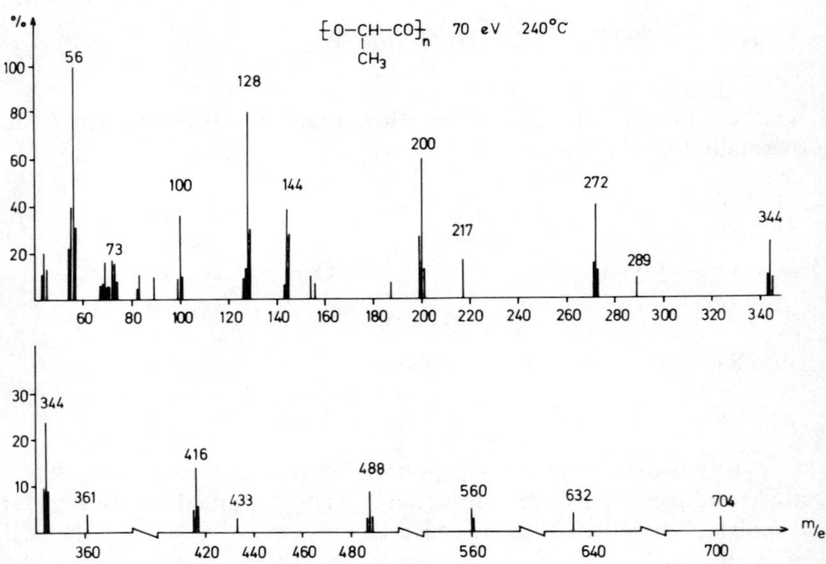

FIG. 4. Pyrolysis-mass spectrum of poly(1,1-oxycarbonylethylene) (**III**) at 240 °C.

The molecular weight of the structural unit of **III** is 72 and the series of peaks at *m/e* 56, 128, 200, 272, 344, 416, 488, 560, 632 and *m/e* 704 (*m/e* 56 + ($n \times 72$)) can be identified as a mixture of oligomeric ions. The smallest of these fragments, *m/e* 56, was proved by a metastable ion at *m/e* 21·5 to derive from the cyclic dimer (*m/e* 144). The fragmentation

† Prepared by cationic polymerisation of the corresponding cyclic dimers.[15]

mechanism shown in eqn. 20 was confirmed by the mass spectrum of an authentic cyclic dimer.

$$\text{cyclic dimer } (m/e\ 144) \xrightarrow[-CO_2]{-CH(CH_3){=}O} \cdot CH(CH_3){-}C{\equiv}O|^+ \ (m/e\ 56) \qquad (20)$$

As a consequence, the higher oligomers of this series, which all contain a radical at the ω-position, can be explained as being derived from a mixture of cyclic oligomers which are subsequently fragmented by the mechanism shown in eqn. 20 (eqn. 21).

$$\cdots O{-}CH(CH_3){-}CO{-}O{-}CH(CH_3){-}CO{-}\cdots \xrightarrow[\text{Second step } -e]{\text{First step } \Delta}$$

$$\text{cyclic oligomer radical cation} \rightarrow \cdot CH(CH_3){-}(CO{-}O{-}CH(CH_3))_x{-}C{\equiv}O|^+ \qquad (21)$$

$m/e(x)$: *56*(0); *128*(1); *200*(2); *272*(3); *344*(4); *416*(5); *488*(6); *560*(7); *632*(8); *704*(9).

A less important reaction, involving thermal cleavage of the ester bond to give linear pyrolysis products is revealed by the corresponding carboxonium ions (eqn. 22).

For both series of oligomeric fragments, m/e $56 + (n \times 72)$ and m/e $73 + (n \times 72)$, an additional cationic depolymerisation was found to occur by following corresponding metastable ion transitions. This reaction corresponds to a reverse cationic polymerisation, with the difference that the active species is not formed by a catalyst but by an electron impact induced fragmentation (eqn. 23).

$$\cdots\text{—}\underset{\underset{CH_3}{|}}{C}(\text{—}\overset{O}{\overset{\|}{C}}\text{—})\text{—}H \cdots O\text{—}\underset{CH_3}{\underset{|}{CH}}\text{—}CO\text{—}O\text{—}\underset{CH_3}{\underset{|}{CH}}\text{—}CO\text{—}O\text{—}\underset{CH_3}{\underset{|}{CH}}\text{—}CO\text{—}\cdots$$

$$\downarrow \Delta$$

$$HO\text{—}\underset{CH_3}{\underset{|}{CH}}\text{—}CO\text{—}O\text{—}\underset{CH_3}{\underset{|}{CH}}\text{—}CO\text{—}O\text{—}\underset{CH_3}{\underset{|}{CH}}\text{—}CO\text{—}\cdots \quad (22)$$

$$\downarrow -e$$

$$H\text{—}(O\text{—}\underset{CH_3}{\underset{|}{CH}}\text{—}CO)_x\text{—}O\text{—}\underset{CH_3}{\underset{|}{CH}}\text{—}C{\equiv}O|^+$$

m/e(*x*): *73*(0); *145*(1); *217*(2); *289*(3); *361*(4); *433*(5).

$$\cdot\underset{CH_3}{\underset{|}{CH}}\text{—}CO\text{—}O\text{—}\underset{CH_3}{\underset{|}{CH}}\text{—}CO\text{—}O\text{—}\underset{CH_3}{\underset{|}{CH}}\text{—}CO\text{—}O\text{—}\underset{CH_3}{\underset{|}{CH}}\text{—}C{\equiv}O|^+ \qquad m/e\text{: }272$$

$$m^*\ 147 \downarrow -\overline{O\text{—}\underset{CH_3}{\underset{|}{CH}}\text{—}CO}$$

$$\cdot\underset{CH_3}{\underset{|}{CH}}\text{—}CO\text{—}O\text{—}\underset{CH_3}{\underset{|}{CH}}\text{—}CO\text{—}O\text{—}\underset{CH_3}{\underset{|}{CH}}\text{—}C{\equiv}O|^+ \qquad m/e\text{: }200$$

$$m^*\ 82 \downarrow -\overline{O\text{—}\underset{CH_3}{\underset{|}{CH}}\text{—}CO} \qquad (23)$$

$$\cdot\underset{CH_3}{\underset{|}{CH}}\text{—}CO\text{—}O\text{—}\underset{CH_3}{\underset{|}{CH}}\text{—}C{\equiv}O|^+ \qquad m/e\text{: }128$$

$$m^*\ 24{\cdot}7 \downarrow -\overline{O\text{—}\underset{CH_3}{\underset{|}{CH}}\text{—}CO}$$

$$\cdot\underset{CH_3}{\underset{|}{CH}}\text{—}C{\equiv}O|^+ \qquad m/e\text{: }56$$

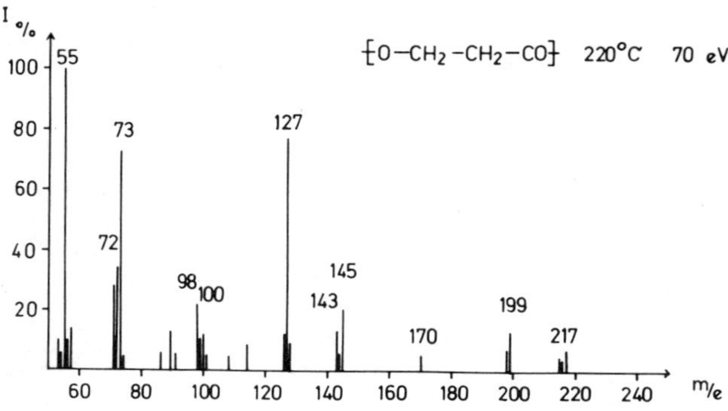

FIG. 5. Pyrolysis-mass spectrum of poly(oxycarbonylethylene) (**IV**) at 220 °C.

In earlier investigations[16,17] acrylic acid was reported as the main pyrolysis product of poly(oxycarbonylethylene) (**IV**). The pyrolysis-mass spectrum of **IV** (Fig. 5) confirms these results in so far as the formation of oligomers of acrylic acid can be deduced by the detection of carboxonium ions formed by subsequent fragmentation (eqn. 24).

$$\cdots\text{—CH}_2\text{—C(=O)—O—}\quad \text{(H—)CH—CO—O—} \cdots\cdots \text{—CH}_2\text{—C(=O)—O}\quad \text{CH}_2\text{—(H—)CH—CO—}\cdots$$

$$\downarrow \Delta$$

$$\text{CH}_2\text{=CH—(CO—O—CH}_2\text{—CH}_2)_x\text{—CO—OH} \qquad (24)$$

$$\downarrow -e$$

$$\text{CH}_2\text{=CH—(CO—O—CH}_2\text{—CH}_2)_x\text{—C}\equiv\text{O}|^+$$

$m/e(x)$: *55*(0); *127*(1); *199*(2).

In addition, the characteristic thermal cleavage of the ester bond is revealed by the detection of the fragments m/e 73, 145 and 217 which occur with small intensities (eqn. 25).

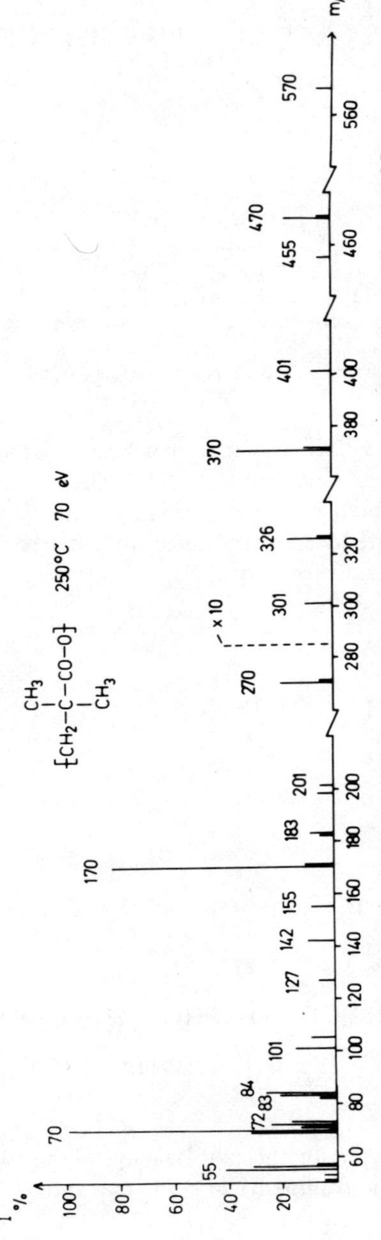

$$\left[CH_2-\underset{CH_3}{\overset{CH_3}{\overset{|}{\underset{|}{C}}}}-CO-O \right]$$

FIG. 6. Pyrolysis-mass spectrum of poly(oxycarbonyl-1,1-dimethyl-ethylene) (V) at 250 °C.

$$\cdots\text{—CH}_2\text{—CH}(\text{H})\text{—}\overset{\text{C=O}}{}\text{O—CH}_2\text{—CH}_2\text{—CO—}\cdots \xrightarrow{\Delta} \text{HO—CH}_2\text{—CH}_2\text{—CO—}\cdots \rightarrow \text{H—(O—CH}_2\text{—CH}_2\text{—CO)}_x\text{—O—CH}_2\text{—CH}_2\text{—C}{\equiv}\text{O}|^+ \qquad (25)$$

$m/e(x)$: *73*(0); *145*(1); *217*(2).

In contrast to poly(oxycarbonylethylene), poly(oxycarbonyl-1,1-dimethylethylene) (**V**, polypivalolactone) cannot undergo a thermal *cis*-elimination. The pyrolysis-mass spectrum (Fig. 6) recorded at 250 °C shows a series of peaks with high intensities at m/e 70 + (n × 100) (MW_{unit} = 100).

By comparison with the mass spectrum of authentic pivalolactone, the fragment m/e 70 could be shown to originate from monomeric pivalolactone by the elimination of formaldehyde (eqn. 26).

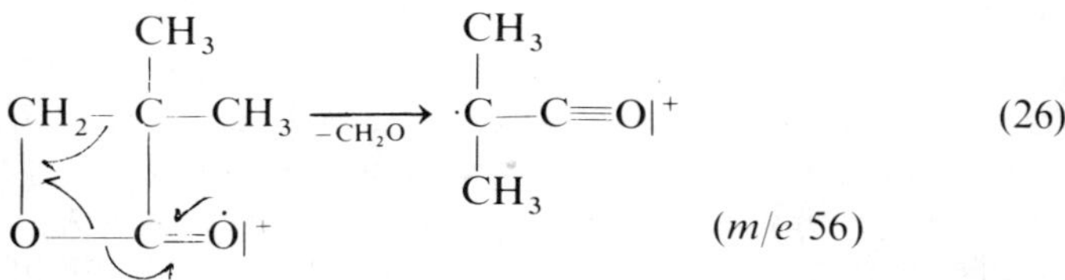

(26)

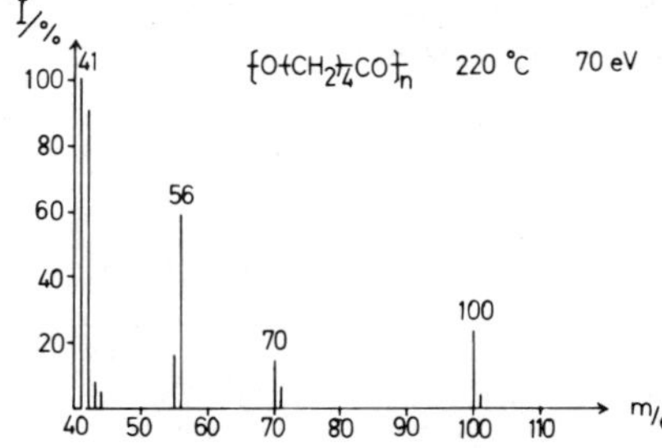

FIG. 7. Pyrolysis-mass spectrum of poly(oxycarbonyltetramethylene) (**VI**) at 220 °C.

This mechanism (eqn. 26) and the results of previous investigations[18,19] explains the origin of the oligomeric fragments by thermal formation of cyclic products and subsequent fragmentation, as shown in eqn. 27.

$$\left(-O-CH_2-\underset{CH_3}{\overset{CH_3}{\underset{|}{\overset{|}{C}}}}-CO- \right)_n \xrightarrow[\text{Second step } -e]{\text{First step } \Delta} \text{cyclic } [O-CH_2-C(CH_3)_2-CO-O\sim\sim C(=O^{\cdot +})] \quad (27)$$

$$\downarrow -CH_2O$$

$$\cdot\underset{CH_3}{\overset{CH_3}{\underset{|}{\overset{|}{C}}}}-\left(CO-O-CH_2-\underset{CH_3}{\overset{CH_3}{\underset{|}{\overset{|}{C}}}}- \right)_x C\equiv O|^+$$

m/e(*x*): *70*(0); *170*(1); *270*(2); *370*(3); *470*(4); *570*(5).

This degradation behaviour was confirmed by a preparative pyrolysis of **V**. After fractionation of the pyrolysis products by gel permeation chromatography, identical cyclic oligomers were found.[20]

The pyrolysis of poly(oxycarbonyltetramethylene) at 220 °C gives quantitative yields of the six-membered ring δ-valerolactone **(VI)** ($MW = 100$, eqn. 28), and the mass spectrum (Fig. 7) shows no pyrolysis products or fragments of higher masses.

$$\cdots-O-(CH_2)_4-CO-O-(CH_2)_4-CO-\cdots \xrightarrow{\Delta} \text{cyclic } [O-C(=O)-CH_2-CH_2-CH_2-CH_2-] \quad (28)$$

Some studies of the thermal degradation of poly(oxycarbonylpentamethylene) (**VII**, polycarprolactone) have been described previously.[21] The polymer was heated for 20 h at 220 °C under a pressure of 80 mm Hg and considerable amounts of caprolactone ($MW = 114$) were identified in the mixture of non-volatile pyrolysis products. However, only an insignificant decrease in the average molecular weight of the remaining polymer was observed. Hence, a two-step degradation mechanism was postulated which begins at the chain ends and proceeds by a fast depolymerisation according to eqn. 29.

$$\cdots\!-\!O\!-\!(CH_2)_5\!-\!CO\!-\!O\!-\!(CH_2)_5\!-\!CO\!-\!OH \xrightarrow{k_1} \cdots\!-\!O\!-\!(CH_2)_5\!-\!CO\!-\!O\!-\!(CH_2)_5\!-\!CO^+ + OH^- \quad (29)$$

$$\cdots\!-\!O\!-\!(CH_2)_5\!-\!CO\!-\!O\!-\!(CH_2)_5\!-\!CO^+ \xrightarrow[k_2 \gg k_1]{k_2} \cdots\!-\!O\!-\!(CH_2)_5\!-\!CO^+ + \underbrace{O\!-\!(CH_2)_5\!-\!CO}_{\text{ring}}$$

In contrast to this assumption is the observation that the pyrolysis-mass spectrum of **VII** (Fig. 8, 220 °C) contains only insignificant amounts of monomeric caprolactone (m/e 114). Instead, a series of oligomeric ions at m/e $115 + (n \times 114)$ is found with high intensities.[22]

Consistent with the degradation behaviour of aliphatic polyesters of dicarboxylic acids and diols, the formation of these fragments can be explained by a thermal cleavage of the ester bond and formation of ketene and hydroxyl end groups (eqn. 30).

$$\cdots\!-\!CH(H)\!-\!C(=O)\!-\!O\!-\!CH_2\!-\!CH_2\!-\!CH_2\!-\!\cdots \xrightarrow{\Delta} \cdots\!-\!CH\!=\!C\!=\!O + HO\!-\!CH_2\!-\!CH_2\!-\!CH_2\!-\!\cdots \quad (30)$$

The electron impact induced fragmentation of these pyrolysis products with ketene and hydroxyl end groups leads to carboxonium ions or the elimination of water yields two series of oligomers with a thermally formed

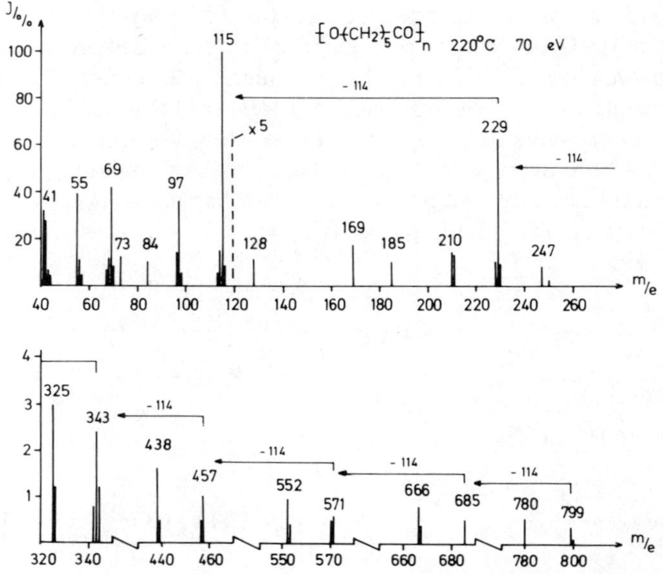

FIG. 8. Pyrolysis-mass spectrum of poly(oxycarbonylpentamethylene) (**VII**) at 220 °C.

hydroxyl (eqn. 31) or a thermally formed ketene end group (eqn. 32).

$$H\text{—}(O\text{—}(CH_2)_5\text{—}CO)\text{—}_xO\text{—}(CH_2)_5\text{—}C{\equiv}O|^+ \quad (31)$$

m/e(*x*): *115*(0); *229*(1); *343*(2); *457*(3); *571*(4); *685*(5); *799*(6).

$$^+CH_2{=}CH\text{—}((CH_2)_3\text{—}CO\text{—}O\text{—}(CH_2)_2)_xCH_2\text{—}CH_2\text{—}CH{=}C{=}O \quad (32)$$

m/e(*x*): *96*(0); *210*(1); *438*(3); *552*(4); *666*(5); *780*(6).

By measurement of the appropriate metastable ion transitions, a subsequent cationic depolymerisation of these carboxonium ions can be established (compare eqn. 23), and the formation of *m/e* 115 by three elimination steps of caprolactone from *m/e* 457 in a time $t < 10^{-5}$ s can be observed. Although this electron impact induced degradation corresponds to the second step of the mechanism postulated in eqn. 29, a final explanation of this discrepancy between the earlier and later results cannot be given. It might be supposed, however, that the degradation into hydroxyl and ketene end groups (eqn. 30) can only be detected in high vacuum, which provides for the fast removal of these reactive species, whereas at higher pressure a back reaction becomes dominating so that only the concurrent degradation to monomer is observed (eqn. 33).

$$
\begin{array}{c}
\text{eqn. 30} \\
\uparrow 10^{-6}\,\text{mm Hg} \\
\cdots\!-\!O\!-\!(CH_2)_4\!-\!CH\!=\!C\!=\!O + HO\!-\!(CH_2)_5\!-\!CO\!-\!\cdots \\
\swarrow\!\nearrow \Delta \\
\cdots\!-\!O\!-\!(CH_2)_5\!-\!CO\!-\!O\!-\!(CH_2)_5\!-\!CO\!-\!\cdots \\
80\,\text{mm Hg} \downarrow \\
\underbrace{O\!-\!(CH_2)_5\!-\!CO}
\end{array} \qquad (33)
$$

CONCLUSION

The favoured pyrolysis mechanisms of different polyesters in high vacuum, shown by experiments of direct degradation in the mass spectrometer, can be summarised as follows:

(1) Polyesters of terephthalic acid and aliphatic diols are preferentially degraded by thermal *cis*-elimination, a rearrangement process which avoids the homolytic cleavage of bonds into radicals. A less important reaction involving thermal cleavage of the ester bond is also observed.

(2) The most favoured pyrolysis reaction of polyesters of aliphatic dicarboxylic acids and diols is the cleavage of the ester bond which leads to volatile products with ketene and hydroxyl end groups. Simultaneously, small amounts of pyrolysis products are detected which derive from a *cis*-elimination or a cleavage inside the dicarboxylic acid units.

(3) The favoured thermal degradation reactions of polyesters of α-hydroxy carboxylic acids or lactones depend strongly on the ring size of the corresponding lactones and their substituents. Whereas poly-β-propiolactone decomposes mainly by thermal *cis*-elimination into linear oligomers of acrylic acid, poly-α-dimethyl-β-propiolactone (polypivalolactone) is unable to undergo this

reaction and yields a mixture of cyclic oligomers. Poly-δ-valerolactone decomposes quantitatively into the six-membered δ-valerolactone and the pyrolysis of poly-ε-caprolactone leads to linear oligomers by cleavage of the ester bond. Polyesters of α-hydroxy acids (polylactic, polyglycolic) degrade thermally in high vacuum into cyclic oligomers, as well as into linear ω-hydroxy ketenes by cleavage of the ester bond.

REFERENCES

1. GRASSIE, N., *Chemistry of High Polymer Degradation Processes*, London, Butterworth, 1956.
2. MADORSKY, S. L., *Thermal Degradation of Organic Polymers*, Polymer Reviews, Vol. 7, New York, Interscience Publ., 1964.
3. JONES, C. E. R. and CRAMERS, C. A., *Analytical Pyrolysis*, Amsterdam, Elsevier Scientific Publishing Company, 1977.
4. WILLIAMS, C. G., *Phil. Trans.*, **150** (1860), p. 241.
5. LÜDERWALD, I., In *Polymer Spectroscopy, II*, Hummel, D. O. (Ed.), pp. 217–55, Weinheim, Verlag Chemie, 1979.
6. SPITELLER, G., *Massenspektrometrische Strukturanalyse Organischer Verbindungen*, S.32, Weinheim, Verlag Chemie, 1966.
7. LÜDERWALD, I., unpublished results.
8. AGUILERA, C. and LÜDERWALD, I., *Makromol. Chem.*, **179** (1978), p. 2817.
9. LÜDERWALD, I. and URRUTIA, H., *Makromol. Chem.*, **177** (1976), p. 2079.
10. LÜDERWALD, I. and URRUTIA, H., In *Analytical Pyrolysis*, Jones, C. E. R. and Cramers, C. A. (Eds.), pp. 139–49, Amsterdam, Elsevier Scientific Publishing Company, 1977.
11. GOODINGS, E. P., *Soc. Chem. Ind.*, Monograph 13 (1961), p. 211.
12. RITCHIE, P. D., *Soc. Chem. Ind.*, Monograph 13 (1961), p. 106.
13. LÜDERWALD, I. and URRUTIA, H., *Makromol. Chem.*, **177** (1976), p. 2093.
14. JACOBI, E., LÜDERWALD, I. and SCHULZ, R. C., *Makromol. Chem.*, **179** (1978), p. 429.
15. JACOBI, E., Dissertation, Mainz, in preparation.
16. GRSEHAM, T. L., JANSEN, J. E. and SHAVER, F. W., *J. Am. Chem. Soc.*, **70** (1948), pp. 998, 1004.
17. IWABUCHI, S., JAACKS, V., GALIL, F. and KERN, W., *Makromol. Chem.*, **165** (1973), p. 59.
18. WILEY, R. H., *Macromolecules*, **4** (1971), p. 254.
19. KRICHELDORF, H. R. and LÜDERWALD, I., *Makromol. Chem.*, **179** (1978), p. 421.
20. LÜDERWALD, I. and SAUER, R., in preparation.
21. IWABUCHI, S., JAACKS, V. and KERN, W., *Makromol. Chem.*, **177** (1976), p. 2675.
22. LÜDERWALD, I., *Makromol. Chem.*, **178** (1977), p. 2603.

Chapter 4

ELECTRON SPIN RESONANCE STUDIES OF THE MECHANICAL DEGRADATION OF POLYMERS

JUNKICHI SOHMA
Hokkaido University, Japan

SUMMARY

It is concluded on the basis of ESR studies that mechano-radicals, which have been produced by mechanical fracture of polypropylene at low temperature, are trapped on the fresh surfaces. The conversions of the mechano-radicals of poly(methylmethacrylate), polyethylene and polypropylene, which are induced by either heat-treatment or mechanical action, are studied by ESR, and the plausible mechanisms for the radical conversions by the mechanical actions are discussed. The mechano-radicals produced by the ultrasonic irradiation of the solutions of several polymers are investigated by using the spin-trapping method and the species of mechano-radicals are identified by ESR spectra. The identifications provide the direct evidence that main-chain scissions are induced by ultrasonic action. It was experimentally verified that the changes in the crystalline structures of polypropylene and polyethylene are induced by the ball-milling of the samples at low temperatures.

1. INTRODUCTION

It has been established by ESR that mechanical fracture of solid polymers produces free-radicals through scission of the polymer chains.[1] Free-radicals formed by mechanical action are appropriately called 'mechano-radicals'. One of the characteristics of mechano-radicals is high reactivity[1] at low temperatures compared with the corresponding radicals formed by ionisation or photo-irradiation. Reactions of polymer radicals in a solid matrix are considered to be diffusion-controlled and the differences in the reactivity of polymer radicals are mainly attributed not to differences in the radical species but to varying accessibility and mobility. Both accessibility and mobility of radicals may be strongly influenced by their environment. Thus, it is of interest to find out whether or not trapping sites of mechano-radicals are different from those of radicals from other sources. ESR studies on the trapping sites of mechano-radicals will be described in the second section of this chapter.

Since, like other radicals, a mechano-radical is an unstable intermediate, it is able to initiate chemical reactions even at low temperatures when it is subjected to external excitations like annealing or photo-illumination of the polymer matrix in which it is formed and trapped. If reactions initiated by the mechano-radicals involve main-chain scissions of the polymer forming the matrix, degradation of the polymer is accelerated by either thermal- or photo-excitations. It is highly likely that mechano-radicals, which have been produced by a strong impact of a ball in a ball-milling apparatus, experience frequent impacts with other balls if milling time is sufficiently long. In such a circumstance the mechanical action is one of the sources of excitation of the mechano-radicals. Conversion of mechano-radicals induced by either thermal or mechanical agitation is very important, not only in the study of the nature of the mechano-radicals, but for an understanding of the degradation of the polymer. Conversion of mechano-radicals will be discussed in the third section of this chapter.

It has been known for a long time that polymer chains in solution are scissioned by ultrasonic irradiation.[2] Thus, ultrasonic irradiation is an efficient method of producing mechano-radicals in solution. However, no ESR study has been performed on mechano-radicals produced in this way because they decay so rapidly that they have not been observed by the conventional ESR technique. The spin-trapping method[3] was successfully applied to studies on mechano-radicals generated by ultrasonics in polymer solutions. Identification of the radicals, which will be described in the fourth section of this chapter, will provide us with direct experimental evidence for the main-chain scissions induced by ultrasonic waves.

It is known that milling and grinding of inorganic materials very often cause changes in their crystalline structures.[4] The effect of milling on such structural changes in polymers will be discussed in the last section of this review.

This review is not intended to be an exhaustive survey of the literature of ESR studies in the mechano-chemistry of polymers but rather to give a brief summary of our efforts and findings in this particular field.

2. TRAPPING SITES OF MECHANO-RADICALS

Formation of mechano-radicals is always accompanied by the creation of fresh surfaces by rupture of the sample. It may therefore be reasonably assumed that the mechano-radicals are formed primarily on the fresh surfaces. However, this assumption does not necessarily mean that the mechano-radicals are trapped on these surfaces, because a polymer radical in a polymer matrix can migrate from one site to another even at low temperatures.[5-7] Experimental evidence is needed to determine whether the mechano-radicals are trapped on fresh surfaces. In order to obtain experimental verification for the trapping sites of mechano-radicals, polypropylene (PP) was chosen as a sample because there have been a number of ESR studies of PP radicals produced both by irradiation[8-11] and mechanically.[12,13]

2.1. Reactivity of the Mechano-radicals

Polypropylene radicals, which had been formed either mechanically or by γ-irradiation in vacuum, were brought into contact with air at 77 K, and the sample was heat-treated in air for 5 min at various temperatures. The

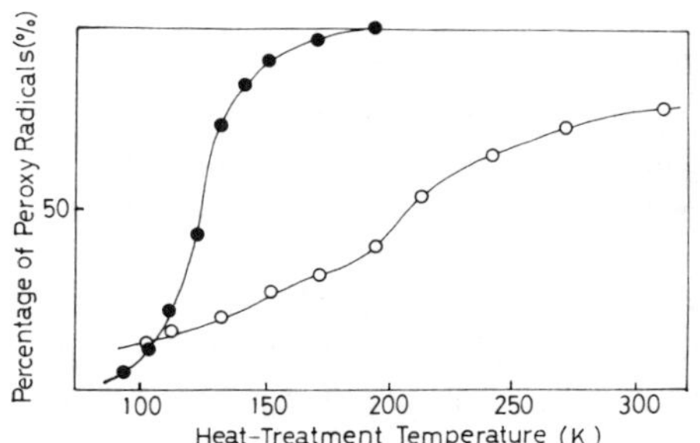

FIG. 1. Reactivity of PP radicals with oxygen: ●; mechano-radical, ○; γ-radical, observed at 77 K.

appearance of peroxyradicals was observed in the spectrum after the heat-treatment.[12-15] The fraction of the peroxyradicals compared with the initial PP-radical concentration was estimated by separation of the spectral components and taking the double integral of each component in the spectrum. This fraction is plotted against temperatures of heat-treatment in Fig. 1. Complete conversion to the peroxyradicals was found to occur at about 200 K in the case of the mechano-radicals, while conversion of radicals produced by γ-irradiation† is about 40 % at this temperature and is never complete even at higher temperatures. This comparison clearly demonstrates the high reactivity of mechano-radicals with oxygen.

Similar experiments were carried out in the presence of methylmethacrylate (MMA) monomer. MMA monomer was introduced at 77 K into the sample tube containing PP mechano-radicals using the special ampoule described by Sohma and Sakaguchi.[1] Figure 2 shows a change in

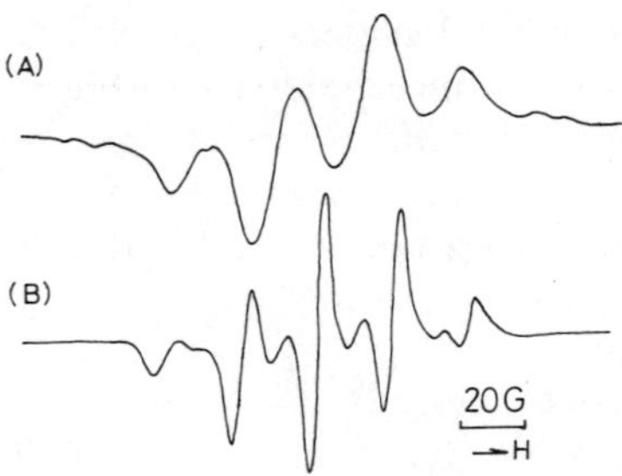

FIG. 2. Comparison of the ESR line-shapes observed from PP mechano-radicals in contact with MMA monomer; (A) before and (B) after heat-treatment at 273 K, 5 min. Observed at 77 K; G = gauss.

the line-shape after heat-treatment at 273 K for 5 min. The line-shape before treatment is the spectrum of the PP mechano-radical[12] while the main part of the spectrum after heat-treatment is the well-known spectrum of poly(methylmethacrylate)-propagating radicals,[16] though a small component of PP mechano-radicals is also present. By contrast with the mechano-radicals, no drastic change in the ESR spectrum of the PP γ-radical was found after similar heat-treatment in the presence of MMA monomer.[15] These experimental results apparently indicate differences in the reactivity between the mechano-radicals and the γ-radicals of PP. It has also been found for other polymers that one of the characteristic features of mechano-radicals is their high reactivity with oxygen molecules.[1]

Mechano-radical and γ-radical species are generally[13,17] different. For example, the PP mechano-radicals are scission types, namely $\sim\!CH_2{-}\dot{C}H(CH_3)$ and $\cdot CH_2{-}CH(CH_3)\!\sim$,[12] and the PP γ-radicals are mostly chain radicals,[8-11] like $\sim\!CH_2{-}\dot{C}(CH_3){-}CH_2\!\sim$. Thus, one might attribute the difference in the reactivity to the different species. However, the accessibility of a reactant to a trapped polymer radical is

† Abbreviated as 'γ-radicals'.

more important in such a reaction, because a reaction in solid polymer is mainly influenced by diffusion.[6] If a reactive free-radical is trapped on a surface, to which any reactant molecules are freely accessible, reaction of the free-radical with the reactant does occur readily. In the case of a small molecule like oxygen, some fraction of oxygen may penetrate into a polymer matrix, in which polymer radicals are trapped, at elevated temperatures, and these penetrating molecules may react with the polymer radicals situated in the matrix. At these temperatures polymer radicals decay through their recombination. In such a case, there is a competition between a reaction with foreign molecules and a decay reaction, and therefore 100% conversion is never obtained. This is the case for the γ-radicals, as shown in Fig. 1. Penetration of such a bulky reactant as MMA into the polymer matrix is hardly expected. Accordingly, the polymer radicals are able to react with the bulky molecules only when they are trapped on the surfaces. The near 100% scavenging of the mechano-radicals by MMA indicates that most of the mechano-radicals are trapped on the surface.

2.2. Decay of the Peroxyradicals

Isothermal decay of the peroxy-mechano-radicals as well as the peroxy-γ-radicals are shown in Fig. 3. There is a clear difference in the decay behaviours of these radicals, depending on the different origins. Approximately 90% of the peroxy-mechano-radicals decay very rapidly, while the fraction of the radicals decaying at a similar rate is only 50% in the case of the peroxy-γ-radicals. Since the fast decay is attributed to the

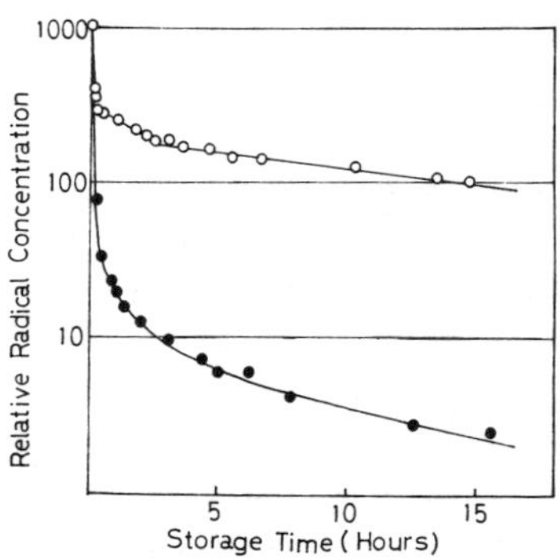

FIG. 3. Isothermal decay of the peroxyradicals at 313 K: ●; mechano-radical, ○; γ-radical.

peroxyradicals trapped in the amorphous regions of solid PP,[18] the experimental result indicates that *ca.* 90% of the mechano-radicals are trapped in amorphous regions. In addition to this difference in the decay behaviour, the line-shape of the peroxy-mechano-radicals at 99 K was considerably different from that of the peroxy-γ-radicals, which showed a typical spectrum of peroxyradicals trapped in a polycrystalline matrix. The spectrum of the peroxy-mechano-radicals was a broad singlet being less asymmetric than the ordinary spectrum of a peroxyradical. This line-shape suggests that the trapped peroxyradicals are sufficiently mobile to average partly the asymmetric spectroscopic splitting factor (g factor) of the peroxyradical. From these experimental results one may conclude that the majority of the peroxy-mechano-radicals are trapped in a region in which both rapid molecular motion and easy decay are permitted. The fraction of the slow decaying peroxy-γ-radicals is approximately equal to the crystallinity of the PP used. The radicals produced by γ-irradiation are formed and trapped homogeneously all over the sample and the percentage of the radicals which are trapped in the crystalline part and decay slowly are proportional to the crystallinity. The fact that the peroxy-mechano-radicals are trapped in an amorphous and mobile phase is consistently interpreted by assuming that more than 90% of mechano-radicals are trapped on surfaces, presumably fresh surfaces produced by fracture.

2.3. Photo-conversion of the Peroxy-mechano-radicals[15]

It was found that mechano-radicals react very efficiently with oxygen adsorbed on the PP to form peroxyradicals. The peroxyradicals formed in such a way were easily decomposed by u.v. illumination, although the peroxy-γ-radicals were not easily photo-decomposed. The easy reactivity with the sorbed oxygen and the rapid photo-conversion also suggest that the mechano-radicals are trapped on the surfaces.

2.4. Trapping Site of the Mechano-radicals

Taken together, the various results described above indicate that the PP mechano-radicals are formed and trapped on the fresh surfaces produced by fracture. Comparison between the trapping sites of the mechano-radicals and the γ-radicals is schematically represented in Fig. 4. These conclusions about PP may be extended more or less similarly to other crystalline polymers[1,19] because similar features, such as high reactivity, are found for mechano-radicals. However, it is worth mentioning that there is experimental evidence which suggests that the mechano-radicals of amorphous *cis*-polybutadiene migrate into the matrix.[20]

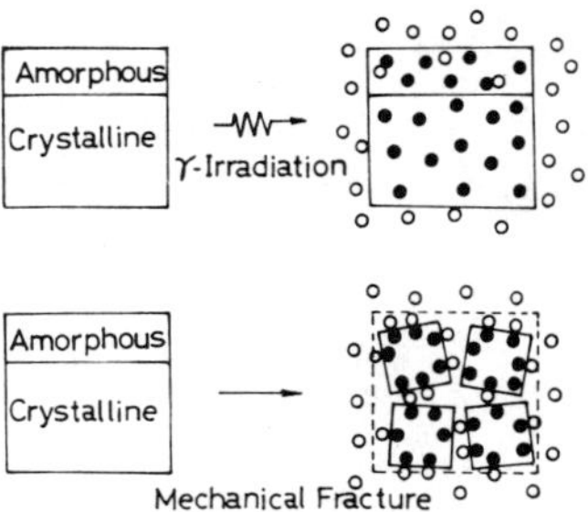

FIG. 4. Schematic pictures of the trapping sites of both the mechano-radicals and the γ-radicals in crystalline polymers: ●; free radicals, ○; adsorbed reactant.

3. CONVERSION OF MECHANO-RADICALS INDUCED BY MECHANICAL ACTION AND HEAT-TREATMENT

3.1. Poly(methylmethacrylate) (PMMA)[21]

3.1.1. *Pair Formation of Mechano-radicals and Conversion by Mechanical Action*

Pair formation of radicals by main-chain scission has been demonstrated for several polymers which had been fractured by ball-milling at 77 K.[1,13] In the case of long milling of PMMA a primary radical of the chain scission was observed but there was also a secondary one.[22] This was interpreted as a rapid conversion of the primary radical. Recently it was found that the ESR spectra changed gradually with the milling time at 77 K in vacuum, as shown in Fig. 5. The spectrum (c) in the figure is apparently separable into a doublet and a nonet, which is the well known spectrum of the scission-type radical (**I**) of PMMA.[23] The doublet is ascribed to the radical (**II**).[22] The spectrum (b) is an intermediate line-shape between the spectrum (a) and (c). The separation of the components of spectrum (a) is shown in Fig. 6.[24] The spectrum consists of three components, a nonet, a doublet, and a triplet. The former two components are identical to those in Fig. 5(c), but the relative intensity of the doublet is much depressed. The triplet, is assigned to radical (**III**). The possibility that the doublet may be attributed to a radical

$$\underset{\textbf{(I)}}{\sim\!\overset{\displaystyle H}{\underset{\displaystyle H}{C}}\!-\!\overset{\displaystyle CH_3}{\underset{\displaystyle R}{C}}\cdot} \qquad \underset{\textbf{(II)}}{\sim\!\overset{\displaystyle CH_3}{\underset{\displaystyle R}{C}}\!-\!\overset{\displaystyle \cdot}{\underset{\displaystyle H}{C}}\!-\!\overset{\displaystyle CH_3}{\underset{\displaystyle R}{C}}\!\sim} \qquad \underset{\textbf{(III)}}{\sim\!\overset{\displaystyle CH_3}{\underset{\displaystyle R}{C}}\!-\!\overset{\displaystyle H}{\underset{\displaystyle H}{C}}\cdot}$$

$$(R = COOCH_3)$$

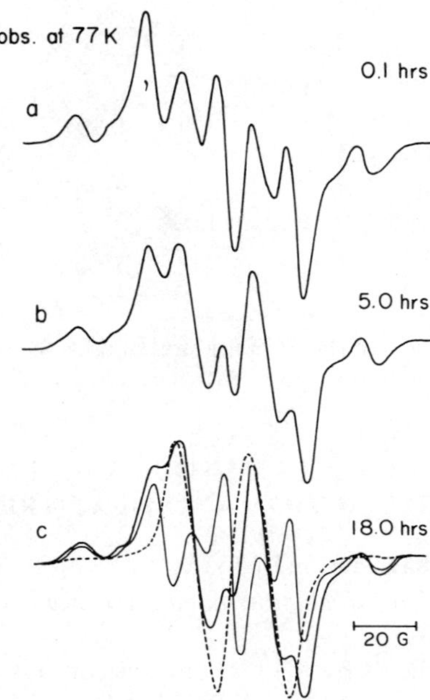

FIG. 5. The ESR spectra observed at 77 K from PMMA in vacuum after various milling times at 77 K.

having a hydrogen-abstracted methyl $\sim CH_2—C(CH_3)(COOCH_2^{\cdot})\sim$ was ruled out by experiments using PMMA with a deuterated methyl-group. It is worthy of note that the sum of the relative intensities of the triplet and the doublet are approximately equal to that of the nonet, which is a primary radical of main-chain scission. Thus, it is reasonable to assume that the triplet originates from the partner radical in the primary pair-formation and the doublet is from the radical (**II**) formed secondarily by hydrogen abstraction of the primary radical (**III**). This spectral change indicates that the radical (**III**), which has been primarily produced by a main-chain scission, is gradually converted to the secondary radical (**II**) as the time of milling at 77 K increases. That is,

$$\underset{\textbf{(III)}}{\sim\overset{\overset{CH_3}{|}}{\underset{\underset{R}{|}}{C}}—CH_2\cdot} \xrightarrow{\text{milling at 77 K}} \sim\overset{\overset{CH_3}{|}}{\underset{\underset{R}{|}}{C}}—CH_3 + \underset{\textbf{(II)}}{\sim\overset{\overset{CH_3}{|}}{\underset{\underset{R}{|}}{C}}—\overset{\overset{H}{|}}{\dot{C}}—\overset{\overset{CH_3}{|}}{\underset{\underset{R}{|}}{C}}\sim} \qquad (1)$$

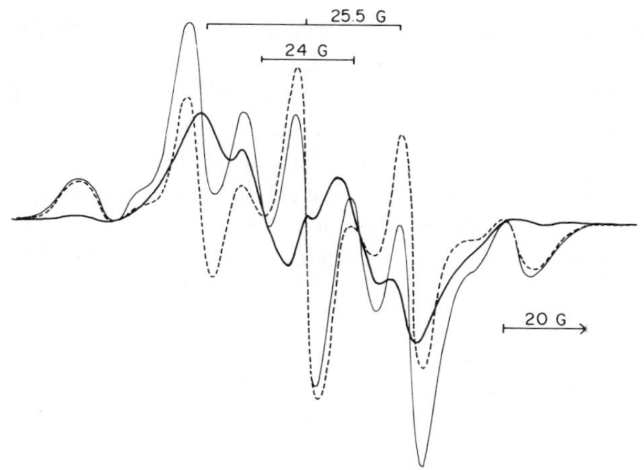

FIG. 6. The thin line is the spectrum observed after short time (0·1 h) milling, at 77 K. The dotted line is the nonet spectrum of the radical (**I**). The bold line is the spectrum, which was obtained by subtraction of the nonet from the thin-line spectrum. This bold-line spectrum consists of a triplet and a doublet, as shown by the stick diagrams. The separation of the doublet is identical to that of the doublet component in Fig. 5(c).

No conversion of the mechano-radicals was observed when the samples containing radicals (**I**), (**II**), and (**III**) were merely stored at 77 K for more than 18 h without any mechanical action. One may conclude from these experimental results that radical conversion (reaction 1) proceeds at a sufficient rate at 77 K only when mechanical action is applied to the system. In other words, it was found that mechanical action induces radical conversion.

3.1.2. *Conversion of the Mechano-radical by Heat-treatment*

After observing a spectrum similar to that in Fig. 5(c) after 18 h milling at 77 K, the sample was heat-treated at various temperatures and the changes in the line-shape were followed. On heat-treatment at 273 K, a gradual change was observed so that the spectrum resembled the line-shape of Fig. 5(a). During this change the total intensity was approximately maintained and the outermost peaks remain unchanged. These facts demonstrate that the nonet, the peaks of which are the outermost, is unchanged but a fraction of the doublet is converted by the heat-treatment into the triplet, which had not been present in the initial spectrum. The resemblance between the spectrum and Fig. 5(a) was so close that the components in both spectra were taken to be identical. Since the triplet originates from the radical (**III**), this spectral change indicates that conversion of the radical (**II**) into the radical (**III**) was induced by the heat-treatment. Although this conversion seems to be a back-reaction of reaction (1), it is unlikely that such a back-reaction is induced by the heat-treatment at 273 K because the radical (**II**) is more stable than the radical (**III**). The most reasonable process for the generation of radical (**III**) from radical (**II**) by heat-treatment is scission of a β-bond from the unpaired electron of the radical (**II**) as follows:

$$\underset{\textbf{(II)}}{\sim\!\begin{array}{c}CH_3\\|\\C\\|\\R\end{array}\!-\!\begin{array}{c}\\ \\ \dot{C}\\|\\H\end{array}\!-\!\begin{array}{c}CH_3\\|\\C\\|\\R\end{array}\!-\left(\begin{array}{c}H\\|\\C\\|\\H\end{array}\!-\!\begin{array}{c}CH_3\\|\\C\\|\\R\end{array}\!-\right)_n} \xrightarrow{\text{heat-treatment}}$$

$$\sim\!\begin{array}{c}CH_3\\|\\C\\|\\R\end{array}\!-CH\!=\!\begin{array}{c}CH_3\\|\\C\\|\\R\end{array} + \underset{\textbf{(III)}}{\cdot\begin{array}{c}H\\|\\C\\|\\H\end{array}\!-\left(\begin{array}{c}CH_3\\|\\C\\|\\R\end{array}\!-\!\begin{array}{c}H\\|\\C\\|\\H\end{array}\!-\right)_n} \quad (n \neq 0) \qquad (2)$$

$$\underset{\textbf{(II}')}{\sim\!\begin{array}{c}H\\|\\C\\|\\H\end{array}\!-\!\begin{array}{c}CH_3\\|\\C\\|\\R\end{array}\!-\!\begin{array}{c}\\ \\ \dot{C}\\|\\H\end{array}\!-\!\begin{array}{c}CH_3\\|\\C\\|\\R\end{array}\!-CH_3} \xrightarrow{\text{heat-treatment}} \underset{\textbf{(III)}}{\sim\!\begin{array}{c}CH_3\\|\\C\\|\\R\end{array}\!-\!\begin{array}{c}H\\|\\C\cdot\\|\\H\end{array}} + \begin{array}{c}CH_3\\|\\C\\|\\R\end{array}\!=\!\begin{array}{c}H\\|\\C\end{array}\!-\!\begin{array}{c}CH_3\\|\\C\\|\\R\end{array}\!-CH_3 \qquad (2')$$

Since the ESR spectra of radicals (**II**) and (**II**′) are identical within the error involved in the experiments, one cannot make a distinction between these radicals from the ESR spectra and one has to assume the presence of both. The reaction schemes (2) and (2′) are degradation processes, which are initiated by mechanical action and accelerated by heat-treatment. The degradation is experimentally supported by a decrease in molecular weight, which was observed after the heat-treatment and is shown as a function of the treatment time in Fig. 7. It is interesting to notice that radical conversion induced by mechanical action is different from that by heat-treatment in the case of PMMA.

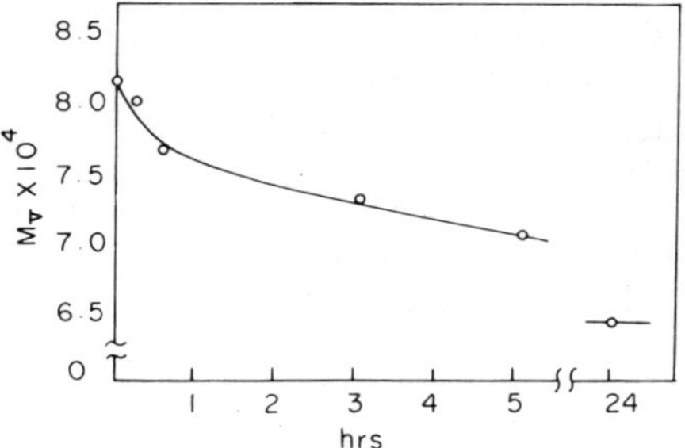

FIG. 7. Decrease in the molecular weight of the fractured PMMA by heat-treatments of various durations at 273 K. The molecular weights were determined by viscosity measurement.

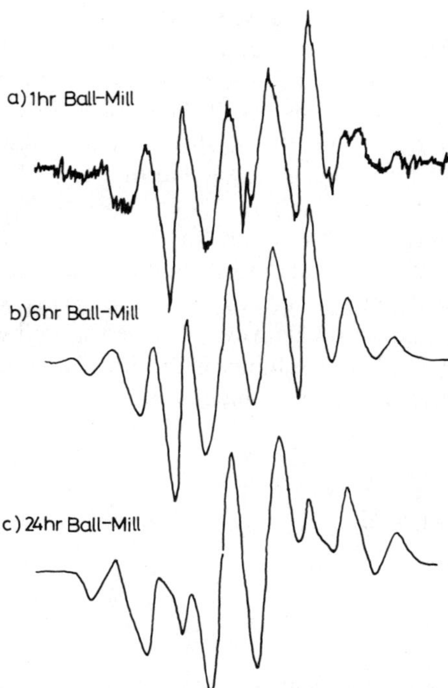

FIG. 8. Variation of the ESR spectra of PE mechano-radicals with milling time observed at 77 K: (a) 1 h milling; (b) 6 h milling; (c) 24 h milling.

3.2. Polyethylene (PE)[25]

The polyethylene (PE) mechano-radical is known to be a scission-type radical (**IV**). This identification was made on the ESR spectrum observed

$$\sim CH_2-\dot{C}H_2$$

(IV)

from PE saw-dusts produced by sawing at 77 K.[13] ESR spectra were observed from PE fractured by the ball-mill at 77 K in vacuum, and examples of the spectra are shown in Fig. 8. The spectrum (a), which was observed after milling for 1 h is the same as the spectrum of the radical $CH_3CH_2CH_2\cdot$ produced by photolysis of $CH_3CH_2CH_2Br$.[26] Accordingly, main-chain scissions of PE were caused by one hour's ball-milling at 77 K. The ESR intensity was increased and the peaks at the wings became clearer after milling for 6 h, as shown in Fig. 8(b). After 24 h milling (Fig. 8(c)), it was found that the relative intensities of the peaks were changed from the original. From these variations of the spectra with milling time one may conclude that the primary mechano-radical of PE is converted to other species at longer milling times. No such changes in the line-shape were observed by keeping the sample, whose spectrum is shown in Fig. 8(a), at 77 K in vacuum for 24 h. This fact means that mechanical action is needed for the conversion of the mechano-radical. Similar changes in the spectrum of the primary mechano-radicals were observed after heat-treatment as shown in Fig. 9. Below 123 K no change in the line-shape was observed. The spectrum (b) in Fig. 9 is an octet, which is assigned to a radical (**V**).[5] The

$$\sim CH_2-\dot{C}H-CH_3$$

(V)

$$\sim CH_2-\dot{C}H-CH_2\sim$$

(VI)

spectrum observed after the heat-treatment at 143 K is noisy due to the radical decay but a number of the peaks apparently decrease. The line-shape appears similar to the sextet of the PE radical (**VI**).[27] These spectral

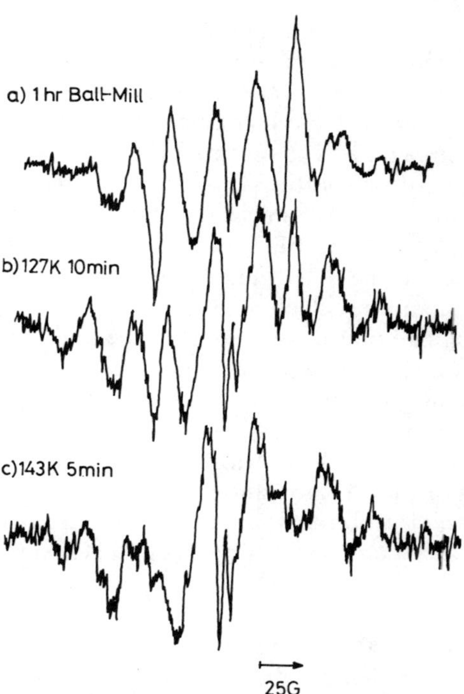

FIG. 9. Changes of the ESR spectrum of PE mechano-radicals after heat-treatment: (a) original spectrum after 1 h milling at 77 K; (b) after heat-treatment at 127 K, 10 min; (c) after heat-treatment at 143 K, 5 min. Observed at 77 K.

changes indicate that the radical species were converted by the heat-treatment as follows:

$$\underset{\textbf{(IV)}}{\sim CH_2-\dot{C}H_2} \xrightarrow[10\,\text{min}]{127\,\text{K}} \underset{\textbf{(V)}}{\sim CH_2-\dot{C}H-CH_2-H} \xrightarrow[5\,\text{min}]{143\,\text{K}} \underset{\textbf{(VI)}}{\sim CH_2-\dot{C}H-CH_2\sim} \qquad (3)$$

Similar hydrogen abstractions caused by heat-treatment were reported in the case of the scission radical produced by photo-illumination.[5]

The assignments of the components appearing in the spectral changes by heat-treatment help us to analyse the changes of the line-shape caused by longer milling (Fig. 8). The new peaks which appear after milling for 6 h are identical to the outermost peaks of the octet from the radical (**V**). Therefore, the spectrum is a mixture of the spectra of the radicals (**IV**) and (**V**). In the spectrum (c) of Fig. 8 the component of the sextet from the radical (**VI**) is rather enhanced, though the octet component from the radical (**V**) is still present. The spectrum is a mixture of the sextet and the octet. Such comparisons of the spectra lead one to believe that the radical conversions induced by mechanical action are similar to those induced by heat-treatment, that is the PE mechano-radical $\sim\!\!\sim CH_2—CH_2\cdot$ (radical (**IV**)) is converted to the other radicals (**V**) and (**VI**) by milling for a long time. The PE sample, which had contained the γ-radical by pre-γ-irradiation, was ball-milled for a longer time and the ESR spectra were observed and compared with the spectral changes of the mechano-radicals. Although the ESR spectrum was changed by milling, the changes observed originated from mixing with the spectrum of the new mechano-radicals and no conversion of the γ-radicals was observed.

Conversions of the PE mechano-radicals were induced by mechanical action and the radical conversions are the same as those induced by heat-treatment. No such conversion by mechanical action was observed for the γ-radicals of PE.

3.3. Polypropylene (PP)[25]

No conversion of mechano-radicals was observed after 24 h of milling of a PP sample which had contained the primary mechano-radicals of PP. It was concluded from the changes in ESR spectra, that radical conversion occurs on heat-treatment but only when the heat-treatment is carried out at temperatures above 273 K.

3.4. Polymerisation Induced by Mechanical Action[19]

It was found that polymerisation was initiated and proceeded at 77 K by milling polytetrafluoroethylene in the presence of monomers such as methylmethacrylate. No polymerisation was initiated at 77 K by introduction of the monomers to the system containing the mechano-radical of polytetrafluoroethylene. Mechanical action was necessary for polymerisation which is a kind of conversion of the mechano-radicals.

3.5. Mechanisms of Radical Conversions by Milling

The examples described above demonstrate that radical conversions of mechano-radicals are induced at 77 K by milling. Apparently the energy needed for the conversion is provided by the applied mechanical energy. A question to be answered is: what is the mechanism through which the mechanical energy of the vibrating balls is transferred to the radicals? This is a difficult question to resolve. However, there seems to be four possible processes of transfer of energy from the balls to the radicals in the fractured polymers.

The first is a direct process. Formation of pairs of mechano-radicals, which involves breaking of a chemical bond, is a kind of chemical reaction caused by mechanical action. Several researchers[17,28–31] have proposed various models for main-chain scission by mechanical action. Most of them assume a concentration of stress on a particular chain, which leads to chain-breaking. This is a good example of the direct process, which means direct transfer of mechanical energy to a particular bond resulting in bond scission. Another possibility for the direct process is energy transfer by phonons. Mechanical forces by vibrating balls in a ball-mill are always impulses which act on the polymers in very short periods. The mechanical impulses produce mechanical vibrations of the lattice, namely phonons, in the applied system, and the shorter the pulse is the higher the overtones are contained in it. Thus, the possibility that high energy phonons, $h\nu$, are produced in the system is not completely ruled out in the case of short pulses. If a single phonon, which is made from a sufficiently high frequency lattice vibration, is larger than the activation energy of the radical conversion and is absorbed by the radical, radical conversion is induced by the direct process of mechanical action, just as a photo-chemical reaction is. If a multiphonon process occurs due to a higher density of phonons at a particular mechano-radical site, the direct process is more likely to occur.

The second process is local heating by friction, which always accompanies mechanical action like milling. Even when the temperature of a bulk sample is kept constant, say at 77 K, during an experiment, small regions of the sample will be heated instantaneously by the friction. The temperature at these particular sites may be substantially higher than the bulk equilibrium temperature. If the temperature is sufficiently high at the site of a trapped radical, radical conversion may occur.

The third process is a reaction induced by excess electric charges produced by triboelectricity.[32] Since mechanical action is accompanied by friction, a polymer to which mechanical action is applied is considered to be electrically charged. Excess electric charges, which means the presence of either electrons or holes in the system, may initiate chemical reactions.

The fourth possibility is reactivity of fresh surfaces. Fracture always produces new and fresh surfaces which are not contaminated with adsorbed molecules. It is known that such surfaces have higher surface energies than ordinary surfaces covered with adsorbed molecules.[33] This excess surface energy might by used for radical conversion. Thus, it might be possible that the mechano-radicals trapped on the fresh surfaces are very reactive by reason of the freshness of the surfaces.

A fifth process is a combination of these four mechanisms.

Excess electrons are donated into a PP matrix by photo-ionisation of tetramethyl phenylene diamine dissolved in the solid polypropylene. Reactions which occur under these conditions are quite different from the radical conversions of the PP radicals.[34] This fact seems to suggest that excess charges are less probable origins of radical conversion. No radical conversion was found when the samples were stored at 77 K without milling. If the freshness of the surfaces is a main cause for the high reactivity of the mechano-radicals, a reaction might be expected to proceed during storage at 77 K, because there must have been some fresh surfaces formed immediately before cessation of the milling. The difference in surface energy between a clean and a contaminated surface is larger in the case of substances having ionic bonds, but one would not expect such a large difference for a non-polar material such as a polymer.[4,33] Thus, it is not completely unreasonable to omit this 'fresh surface mechanism' in the later discussion.

In the case of PE the conversions caused by milling are similar to those by heat-treatment. This similarity strongly suggests that mechanically induced radical conversion involves a process similar to thermal action. For this reason, a plausible mechanism for radical conversion in the case of the PE mechano-radicals is local heating. If so, the local temperatures at the site at which the primary mechano-radicals are converted, is at least 127 K, which is the lowest temperature of the same radical conversion induced by heat-treatment. The reason why no conversion was observed for the PP mechano-radical is probably that the local temperature due to friction is not raised to such a high temperature as the 273 K which is required for this radical conversion. The absence of conversion of the γ-radical may be explained by assuming that local heating instantaneously occurs only on surfaces where frictional forces are acting. Since the mechano-radicals are trapped on the surface as mentioned in Section 2, local heating has an effect only on mechano-radicals but not on γ-radicals, most of which are trapped inside the polymer. These considerations lead us to believe that the experimental results obtained for PE and PP are consistently explained by the local heating process caused by friction.

In the case of PMMA, the mechanical conversion of mechano-radicals is quite different from that by heat-treatment. Heat-treatment causes a self-degradation which consists of successive scissions of the main-chain as mentioned in Section 3.1, while the single step of the hydrogen abstraction is brought about by the mechanical action at 77 K. Since there is no similarity between mechanical and thermal conversion, the direct process might apply for the mechanical conversion of PMMA mechano-radicals. The crudeness of the models and the scarcity of experimental data on radical conversion induced by mechanical action make more definite conclusions unwarranted, so the energy-transfer mechanism in this case is open to debate.

4. APPLICATION OF SPIN TRAPPING TO DEGRADATION OF POLYMERS IN SOLUTION BY ULTRASONIC IRRADIATION

Scission of polymer chains by ultrasonic waves is well known and mechanisms of the chain scission have been discussed in terms of the cavitation which occurs in a liquid exposed to ultrasonic waves.[2] However, no ESR studies of polymer degradation by ultrasonics has been reported to the best of the author's knowledge. In recent years the spin-trapping method[3,35] has been successfully applied to the detection of unstable free-radicals in liquid phases, for example the studies of radicals produced by γ-irradiation of liquid samples.[36] This method may be applicable to detection of polymer radicals produced by ultrasonic irradiation of polymer solutions and results obtained may provide us with interesting information on the degradation of polymers in solution. If one can successfully detect an ESR spectrum from a chain scission radical under ultrasonic irradiation, this is direct experimental evidence for a chain scission caused by ultrasonics.

4.1. The Spin-trapping Method

Nitroxide compounds scavenge unstable radicals to convert themselves into stable radicals, which are called spin-adducts. For example, pentamethylnitrosobenzene (PMNB)[37] attacks an unstable radical, ·R, to form a nitroxide radical.

$$\mathrm{(Me)_5C_6{-}N{=}O} \rightarrow \mathrm{(Me)_5C_6{-}N(\dot{O}){-}R} \qquad (4)$$

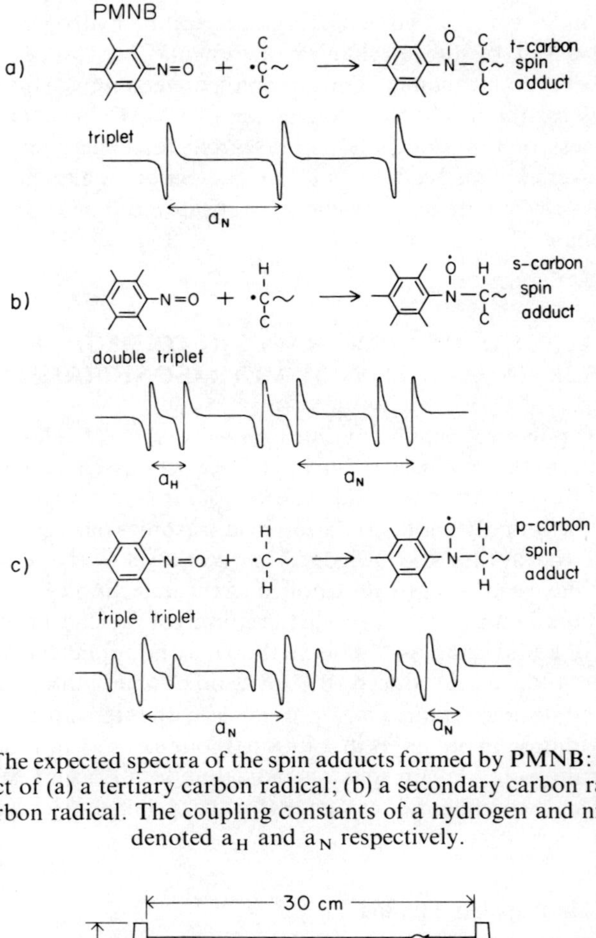

PMNB

a) N=O + •C~ → N–C~ t-carbon spin adduct

triplet

a_N

b) N=O + •C~ → N–C~ s-carbon spin adduct

double triplet

a_H a_N

c) N=O + •C~ → N–C~ p-carbon spin adduct

triple triplet

a_N a_N

FIG. 10. The expected spectra of the spin adducts formed by PMNB: A spectrum of an adduct of (a) a tertiary carbon radical; (b) a secondary carbon radical; (c) a primary carbon radical. The coupling constants of a hydrogen and nitrogen, are denoted a_H and a_N respectively.

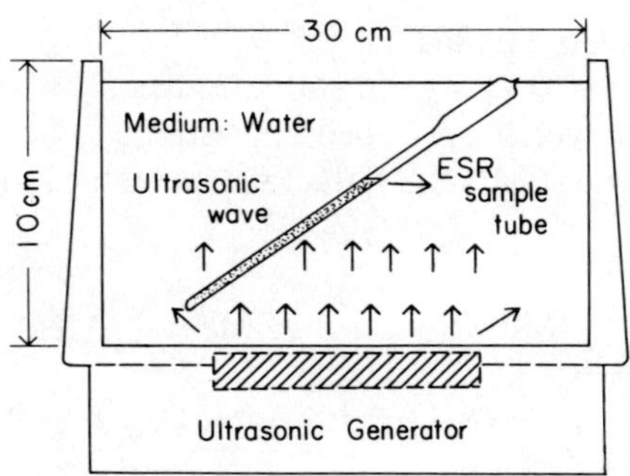

FIG. 11. Schematic diagram of the apparatus for ultrasonic irradiation.

Suppose this unstable radical has an unpaired electron at a primary carbon,† like $\cdot CH_2{-}R'$, an ESR spectrum observed from the spin-adduct

$$\mathrm{(Me)_5C_6{-}N(\dot{O}){-}CH_2{-}R'}$$

is a triple-triplet, in which a main triplet originates from a coupling with the nitrogen and the other triplet having 1:2:1 intensities from a coupling with the two hydrogens, as shown in Fig. 10(c). Similarly, a double-triplet is expected from a spin-adduct of a radical having an unpaired electron at a secondary carbon,‡ and a simple triplet is observed from the spin-adduct of a tertiary carbon radical, as illustrated in Fig. 10(a) and (b). If one observes a triplet like the spectrum (a) in Fig. 10 from a system containing PMNB and polymers after ultrasonic irradiation, it is concluded that unstable tertiary carbon radicals are generated in the liquid phase by ultrasonic irradiation of the system. In such a way unstable radicals are detected by ESR in the liquid phase and radical species are uniquely identified under favourable conditions.

4.2. Experimental Arrangement

An experimental arrangement for ultrasonic irradiation is schematically shown in Fig. 11. A commercial ultrasonic cleaner (frequency 45 KHz, power 100 W) was used as ultrasonic source. An ESR sample tube containing a polymer solution with added spin-trapping agent, PMNB in our experiments, was immersed in water, which is a medium for sound propagation. The temperature of the medium is controlled to within ± 1 K by a Coolnics Circulator made by Komatsu Electronics.

4.3. Identification of Mechano-radicals Produced by Ultrasonic Irradiation of Polymer Solutions

4.3.1. *Polystyrene (PS)–Benzene Solution*[38]

Penta-methylnitrosobenzene (4 mg) was dissolved in 1 ml of a concentrated PS–benzene solution (400 mg PS in 1 ml benzene). The solution was irradiated at 313 K. The ESR spectrum obtained after 1 h irradiation is

† Abbreviated as a 'primary carbon radical'.
‡ Abbreviated as a 'secondary carbon radical'.

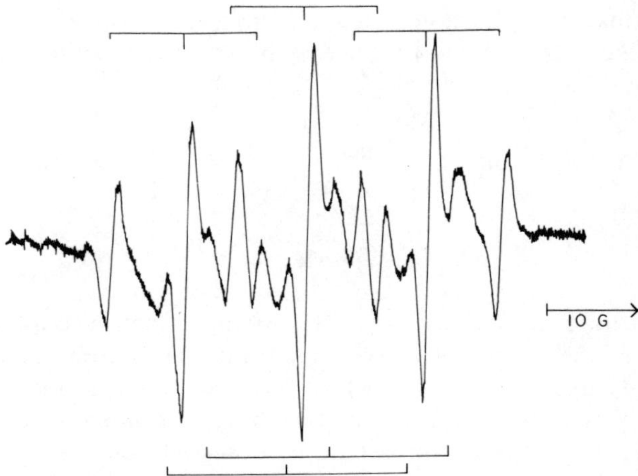

FIG. 12. The spectrum of the PS–PMNB benzene solution after 1 h of ultrasonic irradiation at 313 K. ESR observed at room temperature. The stick diagrams represent the double-triplet and the triple-triplet in Fig. 10.

shown in Fig. 12. The observed spectrum can be separated into two components, a triple-triplet and a triple-doublet as shown by the stick diagrams in Fig. 12. This separation was experimentally supported by the fact that the relative intensities were variable depending upon either irradiation time or storage time. Referring to the spectra in Fig. 10 one can attribute the triple-triplet and the triple-doublet to adducts of a primary carbon radical and of a secondary carbon radical, respectively. Structure (**VII**) is a primary carbon radical originating from a PS mother molecule. There are several secondary carbon radical species possible from PS, namely (**VIII**), (**IX**) and (**X**). The radical species (**IX**) is produced by hydrogen abstraction from a PS molecule. If hydrogen abstractions occur in the system as a secondary reaction, both the radical (**IX**) and another

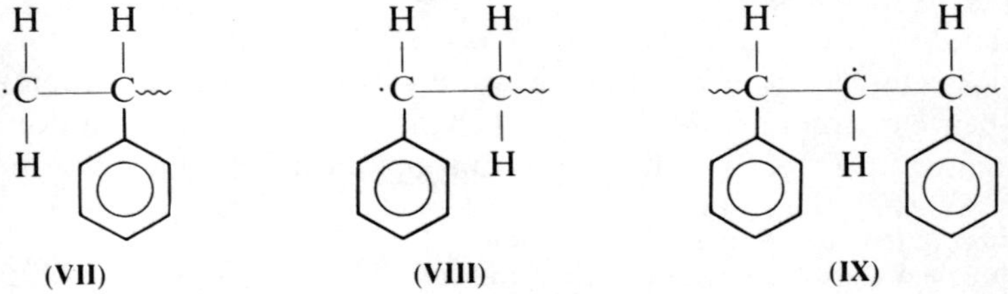

(**VII**) (**VIII**) (**IX**)

species, (**XI**), would be similarly produced. This species (**XI**) is more stable than the others because of the conjugated structure close to the radical so that this radical is surely trapped by PMNB if it exists in the system. No spectral component from an adduct of the tertiary carbon radical (**XI**) was observed. Consequently, no radical species (**XI**) is produced in the system.

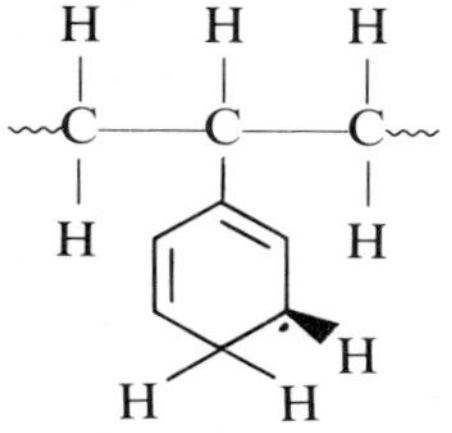

(**X**) 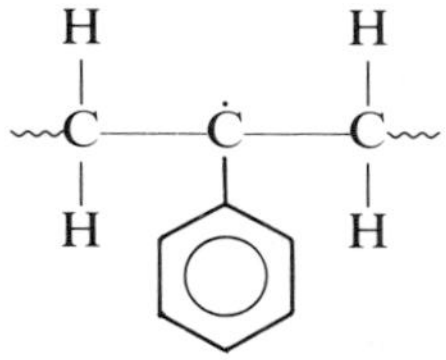 (**XI**)

It follows that the possibility of formation of the radical (**IX**) by hydrogen abstraction is ruled out. Radical (**X**) results from addition of a hydrogen atom which is thus required as a precursor in the system. It is very unlikely that free hydrogen atoms are generated in the system by ultrasonics. Therefore, the existence of the radical (**X**) in the system is highly unlikely. The most plausible structure for the secondary carbon radical is thus (**VIII**), or its resonance structure. The radicals identified by spin-trapping experiments are therefore (**VII**) and (**VIII**), which are a pair species formed by main-chain scission of PS. The main-chain scission of PS in solution by ultrasonics has therefore been experimentally verified by ESR combined with the spin-trapping method.

4.3.2. *Polypropylene–Benzene Solution*[39]

Atactic PP (200 mg) was dissolved in benzene (1 ml) and PMNB (4 mg) was added to the solution. The solution was irradiated at 313 K.

A spectrum observed after 1 h irradiation is shown in Fig. 13. The line-shape of the spectrum might be taken as a single component, triple-triplet. If so, the splitting of the triplet due to the hydrogens is 3·6 gauss (G), which is much smaller than the reported value, 8–11 gauss.[39] Therefore, it is unreasonable to assume that this spectrum is of a single component. However the spectrum can be separated into a triplet and a triple-doublet, as shown by the stick diagrams in Fig. 13. The splitting of the doublet is about 7·2 gauss, which is sufficiently close to the reported value. As a tertiary carbon radical originating from the PP mother molecule, two species, (**XII**) and (**XIII**), are considered possible. As a secondary carbon

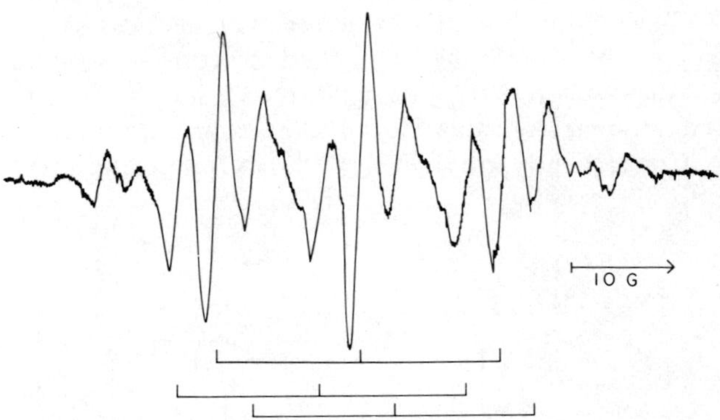

FIG. 13. The spectrum of the PP–PMNB benzene solution after 1 h ultrasonic irradiation at 313 K. ESR observed at the room temperature. The stick diagram represents the triple-triplet in Fig. 10c.

radical from PP one may assume two possible species, **(XIV)** and **(XV)**. There is no additional information at the present stage to allow selection of

H CH_3 H	H CH_3 H	CH_3 H CH_3
H—C—Ċ—C〰	〰C—Ċ—C〰	〰C—Ċ—C〰
H H	H H	H H
(XII)	**(XIII)**	**(XIV)**

one particular species from among these radicals. The species **(XV)** is a radical produced by main-chain scission. The partner of the radical **(XV)** in the chain scission is the radical **(XVI)**. No primary carbon radical was detected by this spin-trapping experiment. However, the radical species **(XII)**, **(XIII)**, and **(XIV)**, which are the species secondarily formed from the other partner **(XVI)** in a chain scission, were also found by the spin-trapping experiments. This fact provides us with good reason to believe that main-chain scissions are caused in PP by ultrasonics.

CH_3 H CH_3	CH_3 H
·C—C—C〰	〰C—C·
H H H	H H
(XV)	**(XVI)**

Similar experiments were carried out on *cis*-polyisoprene, poly(methylmethacrylate) and poly-α-methylstyrene. Analyses of the observed spectra of spin-adducts are now being made.

5. CRYSTALLINE STRUCTURE CHANGES INDUCED BY MECHANICAL FRACTURE[40]

It is known that the crystalline structures of some inorganic materials are changed by milling (for example, see ref. 41). However, the effect of ball-milling on the crystalline structure of solid polymers has not been investigated in detail. It seems of interest to find out whether crystalline structures of polymers are affected by ball-milling at low temperatures. Polypropylene (96 % isotactic) was taken as a sample, on which structural changes after milling were investigated by using ESR, pulsed NMR and X-ray diffraction.

5.1. Effect of Fracture on Radical Conversion

It was shown in Section 2 that PP mechano-radicals are completely converted into PMMA radicals after contact with MMA monomer. The complete conversion of the mechano-radicals indicates that the trapping sites are on the surfaces. Similar experiments on the conversion of γ-radicals were carried out in PP which had been fractured. Polypropylene flakes were fractured in vacuum at 77 K by ball-milling and were sufficiently heat-treated at elevated temperatures in order to destroy all the mechano-radicals. After the disappearance of the mechano-radicals was confirmed by ESR, this sample was γ-irradiated in vacuum and the conversion of the γ-radicals in the sample was carefully studied after contact of the sample with MMA monomer. The ESR spectrum of the PP γ-radicals was considerably changed by the contact with MMA monomer, while little change was observed for the γ-radicals in the non-fractured PP, as mentioned in Section 2. By a separation of the spectrum into its components from PP γ-radicals and from PMMA radicals, the conversion ratio was estimated at about 50 %. This increase in the conversion was apparently due to fracture before γ-irradiation. Since MMA monomers are too bulky to penetrate into the crystallites, the scavenging of the γ-radicals by the MMA monomers must occur on crystallite surfaces to which the MMA monomers are accessible. Therefore, the increase in the conversion from PP γ-radicals to PMMA radicals indicates an increase in the surface-to-volume ratio of the crystallite on an average, assuming uniform formation of radicals in the

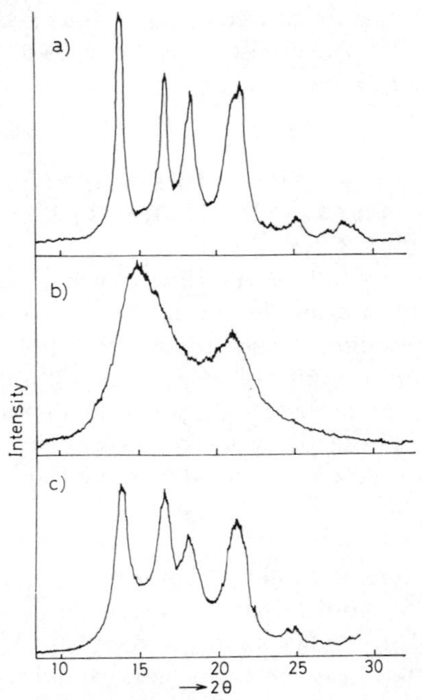

FIG. 14. X-ray diffraction profiles: (a) before and (b) after fracture of PP at 77 K; (c) after heat-treatment of the fractured sample for 92 h at 388 K.

sample by γ-irradiation. Consequently, one is led to believe that the crystallites are fractured and that this decrease in crystallite size is not recovered even by the heat-treatment.

5.2. Changes in X-ray Profiles

X-ray profiles of PP before and after fracture are shown in Fig. 14. The profile before fracture (Fig. 14(a)) is identical to that reported for the monoclinic structure of an isotactic PP.[42] The profile after fracture is quite different from the initial profile and is nearly the same as that of the smectic structure of PP.[42] The changes in X-ray profile are so drastic that it is clear that the crystalline structure of the PP is completely changed from the normal monoclinic structure to the smectic structure by milling for 20 h at 77 K. The pattern (c) in Fig. 14 is the profile obtained after heat-treatment (92 h, 388 K), of the fractured sample which gives the profile Fig. 14(b). The peak positions of the profile coincide with those before fracture, but the

TABLE 1
CHANGES IN CRYSTALLINITY AND IN CRYSTALLITE SIZE BY MILLING

	Crystallinity	*Crystallite size*
Non-fractured PP	71 %	13·7 nm
Fractured PP[a]	53 %	9·7 nm

[a] After heat-treatment at 388 K for 92 h.

diffuse component still remains and the width of each peak is broadened after the heat-treatment. This fact demonstrates that the unit cell structure is reversibly recovered by the heat-treatment, but the crystallinity is not completely recovered. The crystallinity is estimated by the separation of the profiles in Fig. 14(c) into the diffuse and the sharp components, and results are compared in Table 1. The decrease in the crystallinity after fracture was found to be about 18 %. In Fig. 15 the crystallinity obtained from the X-ray

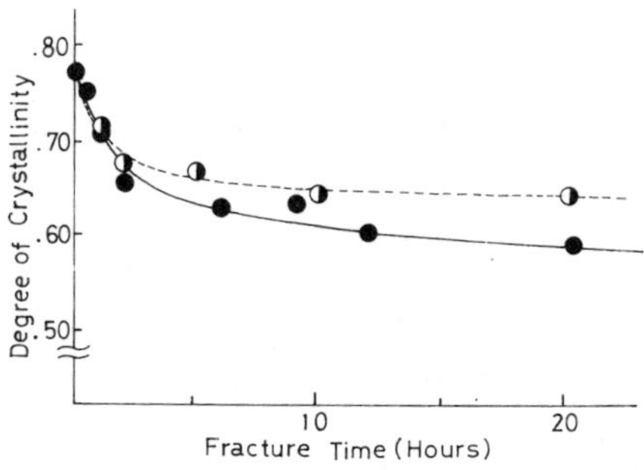

FIG. 15. Milling time dependences of the decrease in crystallinity: ●, PE fractured at 77 K; ◑, PE fractured at room temperature.

profiles of PE are plotted against the milling time. Within a few hours of milling the crystallinity drops sharply and levels off. No marked change was observed by changing the milling temperature.

The other irreversible change was the line-width, which is affected either by crystallite size or by disorders in the crystallites. As mentioned in the last section, the surface increase after fracture is in accord with the decrease in the crystallite size, and it therefore seems reasonable to assume that the line-broadening originates from the decrease in the crystallite size. On this assumption the crystallite size was quantitatively determined by the use of Scherrer's equation:[43]

$$t_{hkl} = \frac{k\lambda}{\beta \cos \theta}$$

where t_{hkl} represents the crystal size in the (hkl) plane; k is a correcting parameter taken as 1·05; λ, the wave length of the X-rays ($\lambda = 0{\cdot}151$ nm); β, the half-width and θ, the diffraction angle. It was found that the crystallite size was decreased to 9·7 nm by milling at the low temperature (Table 1).

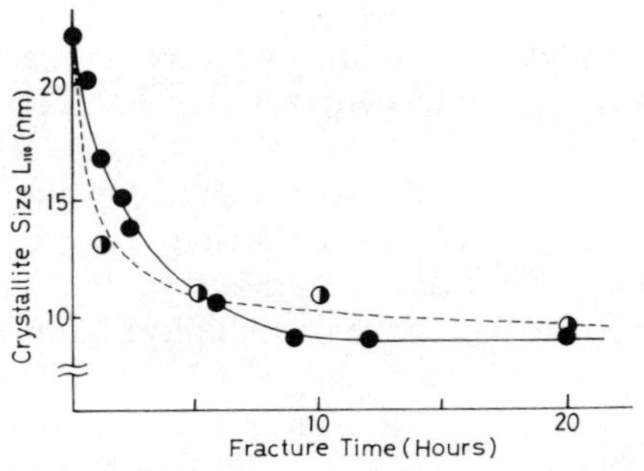

FIG. 16. Milling time dependences of the decrease in crystallinity: ●, PE fractured at 77 K; ◑, PE fractured at the room temperature.

The crystallite sizes of PE estimated by the same method are plotted as a function of the milling time in Fig. 16. In the early stage the crystallite size decreases rapidly to a constant value. Although changes in both crystallinity and crystallite size are illustrated for PE, similar trends were observed for PP.

The conclusions which one may derive from the changes in the X-ray profiles are that crystalline transformation is induced reversibly by mechanical fracture and both crystallinity and crystallite size are decreased irreversibly by fracture. The densities were measured for the three different samples, the non-fractured, the fractured, and the fractured and heat-treated. The observed values are the highest for the non-fractured and the lowest for the fractured. This result supports the suggestion that crystal transformation is induced by fracture.[40]

5.3. Pulsed NMR

A decrease in crystallinity in a solid polymer is equivalent to an increase in the mobile fraction of the polymer. The mobile fraction is easily determined

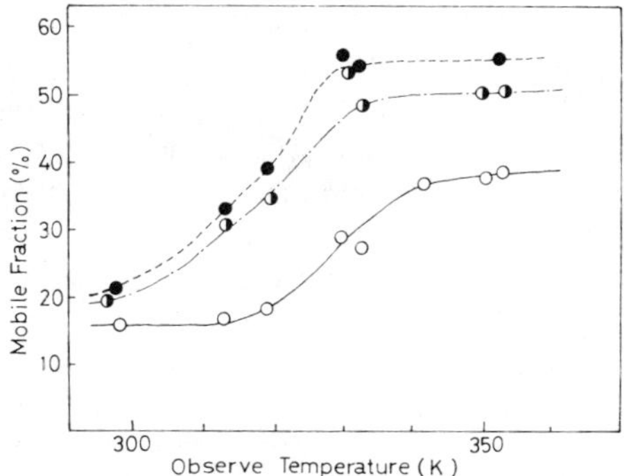

FIG. 17. Temperature dependences of the mobile fraction of PP: ○, non-fractured; ●, fractured for 20 h; ◑, fractured after heat-treatment at 481 K for 24 h.

by measurement of the spin–spin relaxation time, T_2, which is accurately measured by pulsed NMR. T_2 is determined as the gradient of a semilog plot of NMR peak heights, which is expected to be linear. Apparently the plot of the non-fractured sample is represented by the superposition of two straight lines.[44] The curve in the shorter time region is not linear with time but rather with the square of the time. The component of longer T_2 is the mobile fraction. The mobile fractions obtained by this method are plotted against temperature in Fig. 17. Apparently, the mobile fraction of the fractured sample increases at 315 K, at which the mobile fraction of the non-fractured sample stays constant. At higher temperatures the mobile fraction of the fractured sample is larger than that of the non-fractured sample. Since the behaviour of the sample after heat-treatment resembles that of the fractured sample, the increase in the mobile fraction was considered to be irreversible even after heat-treatment. The increase in the mobile fraction caused by the fracture is consistent with the decrease in the crystallinity, which has been deduced from the X-ray experiments.

5.4. Crystalline Structure-changes by Fracture

On the basis of the results obtained by the three different kinds of experiments, radical conversions, X-ray diffraction and T_2-measurements, the following conclusions are consistently drawn:

(1) Fracture of PP by ball-milling at 77 K causes a drastic change in the crystalline structure; fracture induces transformation from the monoclinic to the smectic crystal structure. Bearing in mind the

report[45] that crystal transformation is observed under conditions of large deformation of the bulk sample, one may imagine that the impulses of the milling balls produce a similar effect on the crystalline structure as that of large deformation.

(2) The changes in the crystalline structure of the unit cell are reversible. The unstable smectic structure induced by fracture is restored to the stable monoclinic structure by heat-treatment of the fractured sample.

(3) Decreases in both crystallinity and crystallite size are observed after fracture. The crystallinity and the crystallite size of the fractured sample were never restored to the original values even after the heat-treatment.

ACKNOWLEDGEMENT

The author should like to express his gratitude to Dr M. Tabata, Messrs N. Kurokawa and T. Kudo for valuable discussions and assistance in preparing the figures.

REFERENCES

1. (a) SOHMA, J. and SAKAGUCHI, M., *Adv. Polym. Sci.*, **20** (1976), p. 111.
   (b) ANDREWS, E. A. and REED, P. E., *Adv. Polym. Sci.*, **27** (1978), p. 1.
2. (a) JELLINEK, H. H. G., *Degradation of Vinyl Polymers*, New York, Academic Press, 1955, Chapter 4.
   (b) BASEDOW, A. M. and EBERT, K. H., *Adv. Polym. Sci.*, **22** (1977), p. 83.
3. JANZEN, E. G., *Acc. Chem. Res.*, **4** (1971), p. 31.
4. KUBO, K., *Mekano Kemistori Gairon* (in Japanese), (*Introduction to Mechano Chemistry*), 2nd ed., Tokyo, Tokyo Kagaku-dojin, 1978, Chapter 5.
5. SHIMADA, S., KASHIWABARA, H. and SOHMA, J., *J. Polym. Sci., Part A*-2, **8** (1970), p. 1291.
6. KINELL, P. and RÅNBY, B., *ESR Applications to Polymer Research*, Nobel Symposium 22, 1973, pp. 275, 288.
7. NAGAMURA, T., KUSUMOTO, N. and TAKAYANAGI, M., *J. Polym. Sci., Polym. Phys. Ed.*, **11** (1973), p. 2375.
8. FISCHER, H. and HELLWEGE, K. H., *J. Polym. Sci.*, **56** (1962), p. 33.
9. YOSHIDA, H. and RÅNBY, B., *Acta Chem. Scan.*, **19** (1965), p. 72.
10. TSUJI, K., *Adv. Polym. Sci.*, **12** (1973), p. 132.
11. HAMA, Y., OOI, T., SHIOTSUBO, M. and SHINOHARA, K., *Polymer*, **15** (1974), p. 787.
12. SAKAGUCHI, M., YAMAKAWA, H. and SOHMA, J., *Polym. Letters*, **12** (1974), p. 193.
13. SAKAGUCHI, M. and SOHMA, J., *J. Polym. Sci., Part A*-2, **13** (1975), p. 1233.

14. Fischer, H., Hellwege, K. H. and Neudorff, P., *J. Polym. Sci., Part A*, **1** (1963), p. 2109.
15. Kurokawa, N., Sakaguchi, M. and Sohma, J., *Polym. J.*, **16** (1978), p. 93.
16. Abraham, R. J., Melville, H. K., Ovenall, D. H. and Whiffen, D. H., *Trans. Faraday Soc.*, **54** (1958), p. 133.
17. Kawashima, T., Shimada, S., Kashiwabara, H., and Sohma, J., *Polym. J.*, **5** (1973), p. 135.
18. Eda, B. and Iwasaki, M., *Polym. Letters*, **7** (1969), p. 91.
19. Sakaguchi, M. and Sohma, J., *J. Appl. Polym. Sci.*, **22** (1978), p. 2915.
20. Yamamoto, M., Sakaguchi, M., Shiotani, M. and Sohma, J., *Rept. Prog. Polym. Phys. Japan*, **16** (1973), p. 549.
21. Tabata, M., Yamakawa, H., Takahashi, K. and Sohma, J., *Polym. Degrad. Stability*, **1** (1979), p. 57.
22. Sakaguchi, M., Kodama, S., Edlund, O. and Sohma, J., *J. Polym. Sci., Polym. Letters*, **12** (1974), p. 609.
23. Fischer, H., *Polym. Letters*, **2** (1964), p. 529.
24. Yamakawa, H., Sakaguchi, M. and Sohma, J., *Rept. Prog. Polym. Phys. Japan*, **19** (1976), p. 477.
25. Sohma, J. and Kudo, T., The 27th Polymer Symposium, (Kyoto), The Soc. of Polym. Sci. Japan, 1978.
26. Ayscough, P. B. and Thomson, C., *Trans. Faraday Soc.*, **A-58** (1962), p. 1477.
27. Nara, S., Shimada, S., Kashiwabara, H. and Sohma, J., *J. Polym. Sci., Part A*-2, **6** (1968), p. 1435.
28. Shurkov, S. N., Zakrevski, S. N., Karsukov, V. E. and Kulsenko, V. S., *J. Polym. Sci., Part A*-2, **10** (1972), p. 1509.
29. Peterlin, A., *J. Polym. Sci. C*, No. 32 (1970), p. 297.
30. Kausch, H. H., *Macromol. Sci. Macromol. Chem. C*, **4** (1970), p. 243.
31. DeVries, K. L. and Backman, D. K., *J. Polym. Sci., Part A*-1, **7** (1969), p. 2134.
32. Sakaguchi, M. and Sohma, J., *Polym. J.*, **7** (1975), p. 490.
33. Gaines, G. I. and Tabor, D., *Nature (London)*, **178** (1956), p. 1304.
34. Tabata, M. and Sohma, J., *Polym. Degrad. Stability*, **1** (1979), p. 139.
35. Lagercrantz, C., *J. Phys. Chem.*, **75** (1971), p. 3466.
36. Shiotani, M., Murabayashi, S. and Sohma, J., *Int. J. Radiation Phys. Chem.*, **8** (1976), p. 483.
37. Doba, T., Ichikawa, T. and Yoshida, H., *Bull. Chem. Soc. Japan*, **50** (1977), p. 5158.
38. Miyazawa, T., Tabata, M. and Sohma, J., *Rept. Prog. Polym. Phys. Japan*, **21** (1978), p. 427.
39. Terabe, S., Kuruma, K. and Konaka, R., *J. Chem. Soc., Perkin Trans. II*, (1972), p. 1252.
40. Kurokawa, N. and Sohma, J., *Polym. J.*, in press.
41. Burns, J. H. and Bredig, M. A., *J. Chem. Phys.*, **25** (1956), p. 1281.
42. Natta, G., *Makromol. Chem.*, **35** (1959), p. 94.
43. Alexander, L. E., *X-ray Diffraction Methods in Polymer Science*, New York, John Wiley & Sons, Inc., 1969, Chapter 7.
44. Fujimoto, K., Nishi, T. and Kado, R., *Polymer J.*, **3** (1972), p. 448.
45. Nakamura, K., Inada, K. and Takayanagi, M., *Polymer*, **15** (1974), p. 446.

Chapter 5

THE RÔLE OF LUMINESCENT SPECIES IN THE PHOTOOXIDATION OF COMMERCIAL POLYMERS

NORMAN S. ALLEN† and JOHN F. MCKELLAR‡
University of Salford, UK

SUMMARY

The technique of spectrofluorimetry and phosphorimetry (SFP) although now widely used in other fields is also gaining interest from the polymer chemist. Recent work in our laboratory has clearly demonstrated that polymers such as the polyolefins and aliphatic polyamides contain luminescent (fluorescent and phosphorescent) impurity species which absorb light in the wavelength region 290–350 nm which is most harmful to the polymer. Here the evidence for the identification of the luminescent species in a number of commercial polymers and their behaviour during photo-oxidation is described.

1. INTRODUCTION

Since the early studies of Charlesby and his co-workers[1,2] there has been considerable interest in the luminescence (fluorescence and phosphorescence) from commercial polymers, particularly with regard to the mechanisms of photooxidation and stabilisation. It is the object of this review to describe recent contributions made by the authors.

The most convenient method of classifying the luminescence properties of polymers is that developed by Somersall and Guillet.[3] They classified polymers into two main types. The first, termed type A, emits light from

† Present address: Manchester Polytechnic, UK.
‡ John F. McKellar died suddenly on 3rd May, 1979, after a long spell of illness.

impurity chromophores which do not form part of the molecular structure of the polymer. The second, termed type B, emits light from chromophores which form part of the molecular structure of the polymer. Typical structures for type A and B polymers are:

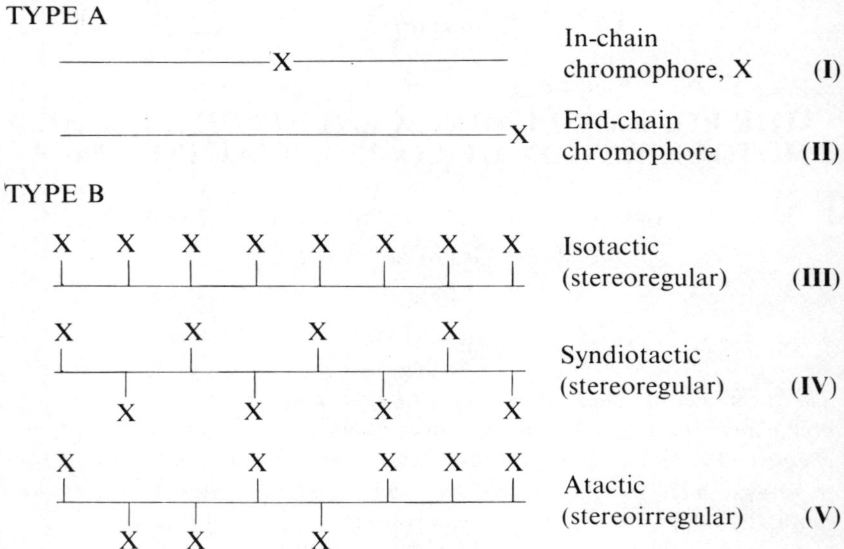

Examples of type A polymers include the polyolefins and aliphatic polyamides in both of which the luminescent chromophore X has been recently identified as an α,β-unsaturated carbonyl group.[4–14] Examples of type B polymers, on the other hand, include aromatic polyesters and polyethersulphones in which the luminescent chromophores have been identified as the aromatic ester carbonyl[15–17] and sulphonyl groups[18,19] respectively. In this chapter the nature of the luminescent chromophores in various type A and B polymers and their rôle during photooxidation is described.

2. TYPE A POLYMERS

2.1. Polyolefins

2.1.1. *Identification of the Luminescent Species*

Figure 1 shows that the fluorescence excitation spectra of the commercial polyolefins, low density polyethylene, polypropylene and poly(4-

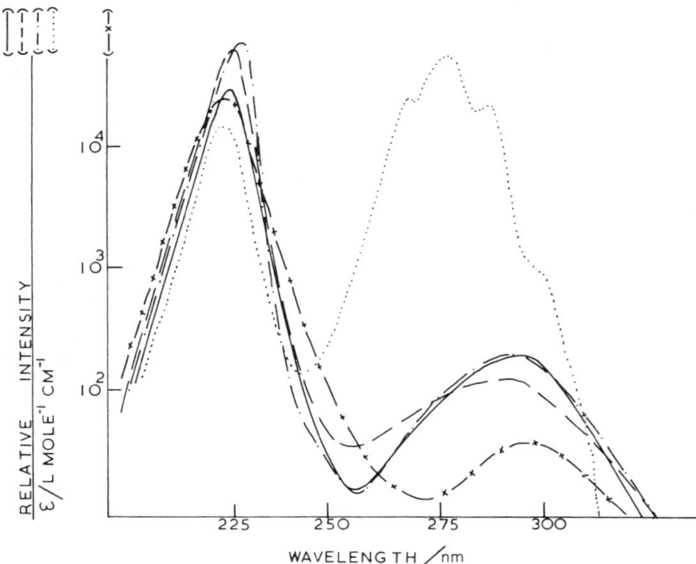

FIG. 1. Comparison of the fluorescence excitation spectra of polypropylene (——), poly(4-methylpent-1-ene) (– – –) and low density polyethylene (– · – ·) films with the absorption spectrum of pent-3-ene-2-one (– × – ×) and the fluorescence excitation spectrum of naphthalene (· · · · ·) in *n*-hexane (10^{-5} M). (Reproduced from *J. Appl. Polym. Sci.*, **21** (1977) pp. 2261, 3147, with permission from John Wiley & Sons Inc., New York.)

methylpent-1-ene) readily match the absorption spectrum of a typical aliphatic enone. The figure also shows that the fluorescence cannot be due to the presence of polynuclear aromatic hydrocarbons, such as naphthalene, in the polymer, which had been postulated by earlier workers.[1,2,20,21] It is seen from Fig. 2 that the excitation spectra differ significantly from that of a fully saturated aldehyde or ketone.

Figure 2 shows that the phosphorescence excitation spectra of the polyolefins match the absorption spectrum of a typical dienone (or–al). It is also seen from the figure that the phosphorescence from the polymers cannot be due to the presence of a typical long-chain aliphatic aldehyde or ketone. In fact, Charlesby and Partridge[1] pointed out that an isolated ketonic carbonyl group on an 'infinite polymethylene chain' may be virtually non-luminescent. A further feature of the phosphorescence emissions from the polymers is that they are long-lived. For low density polyethylene, polypropylene and poly(4-methylpent-1-ene) the emission

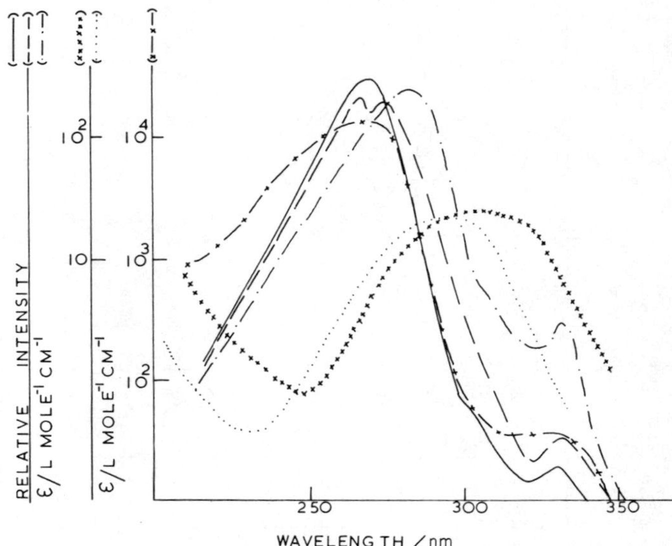

FIG. 2. Comparison of the phosphorescence excitation spectra of polypropylene (——), poly(4-methylpent-1-ene) (– – –) and low density polyethylene (– · – ·) films with the absorption spectra of *trans*, *trans*-hexa-2,4-dienal (– × – ×), 2,2,4,4-tetramethylpentan-3-one (· · · ·) and 3-methylpentanal (× × ×) in *n*-hexane. (Reproduced from *J. Appl. Polym. Sci.*, **21** (1977), pp. 2261, 3147, with permission from John Wiley & Sons Inc., New York.)

lifetimes are about 2·3, 1·6 and 0·7 s respectively.[6–8] Long-lived phosphorescence emission (mean exponential lifetime ($\tau_{1/e}$) > 0·1 s) from rigid, cyclic α,β-unsaturated carbonyl compounds is not uncommon[22–24] and it is possible for such groups to exist as units held within the molecular backbone of the polymer. Further, when present in the polymer matrix, deactivation of these groups by rotation round the C=C bond will be inhibited resulting in a relatively long-lived excited triplet state.

2.1.2. *Effect of Photooxidation on the Luminescent Species*

Figures 1 and 2 show that both the enone and dienone (or–al) chromophoric impurities absorb light in the spectral region 300–350 nm, the region known to be harmful to the commercial polyolefins.

During irradiation under sunlight-simulated conditions the fluorescent enones are gradually consumed.[6–11] This effect is demonstrated in Fig. 3 for polypropylene film where it is seen that there is a gradual reduction in the intensity of the fluorescence excitation spectrum over an irradiation

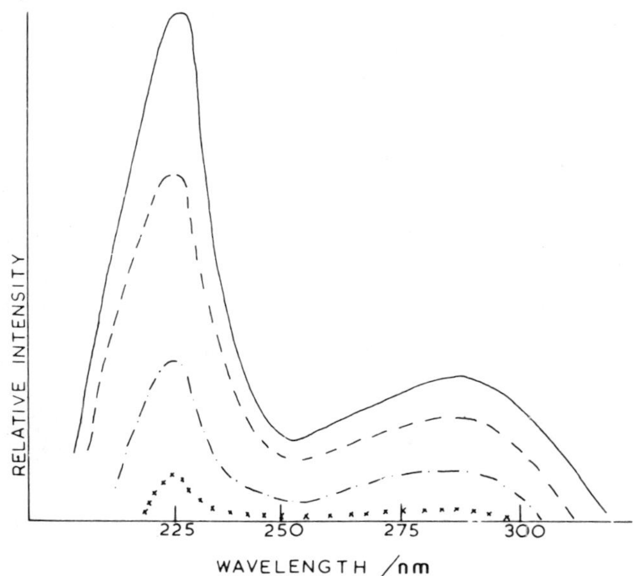

FIG. 3. Intensity of the fluorescence excitation spectrum of polypropylene film before (——) and after 75 h (– – –), 150 h (– · – ·) and 250 h (× × ×) irradiation in a Xenotest-150. (Reproduced from *J. Appl. Polym. Sci.*, **21** (1977), p. 2261, with permission from John Wiley & Sons, Inc., New York.)

period of 250 h. Interestingly, during the first hour there is an initial rapid decrease in the intensity of the fluorescence emission. After this period the rate slows down to a steady value. This effect is clearly demonstrated in Fig. 4 and is probably due to uptake of dissolved oxygen present in the bulk of the polymer film.

The phosphorescent dienones are consumed during irradiation.[6–9,11] Also, at the onset of embrittlement a bathochromic shift in the phosphorescence excitation wavelength maximum is observed. Figure 5 shows this effect for polypropylene film. This shift is due to the conversion of the dienones to saturated carbonyl groups which absorb at longer wavelengths and have a much smaller extinction coefficient.[6] The fluorescence excitation spectrum exhibits no corresponding shift since the saturated aliphatic carbonyl groups formed do not fluoresce.[6]

2.1.3. *Photooxidation Mechanism Involving Luminescent Species*

The photochemistry of α,β-unsaturated carbonyl compounds is well

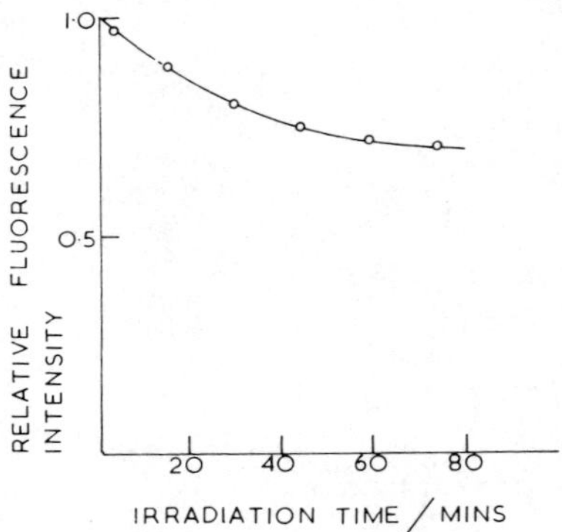

FIG. 4. Effect of irradiation in a Xenotest-150 on the intensity of the fluorescence emission ($\lambda_{max} = 350$ nm) from commercial, unstabilised polypropylene film. (Reproduced from *Polymer*, **18** (1977), p. 968, with permission from IPC Business Press Ltd, UK.)

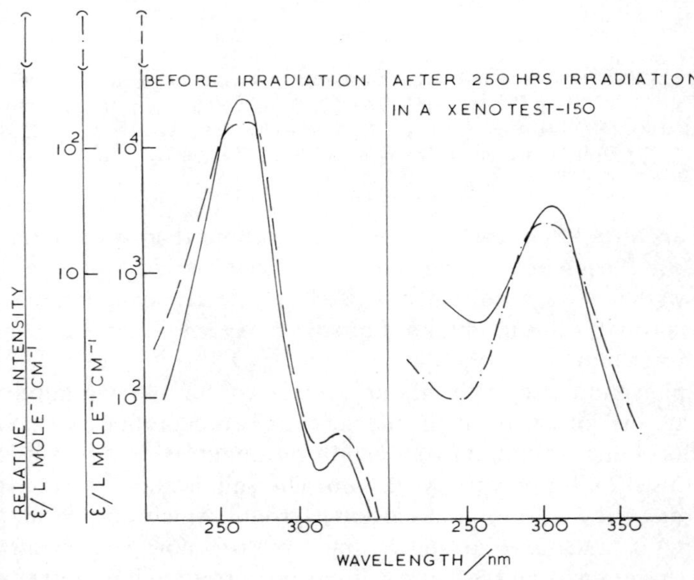

FIG. 5. Comparison of the phosphorescence excitation spectrum of polypropylene film (——) before and after irradiation for 250 h in a Xenotest-150 with the absorption spectra of *trans*, *trans*-hexan-2,4-dienal (– – –) and 2,2,4,4-tetramethylpentan-3-one (– · – ·) in *n*-hexane respectively. (Reproduced from *J. Appl. Polym. Sci.*, **21** (1977), p. 2261, with permission from John Wiley & Sons, Inc., New York.)

documented in the literature.[25,26] These chromophoric groups may undergo two possible photoreactions in the polymer. Considering the enone case for simplicity, these are:

(1) The formation of β,γ-unsaturated carbonyls followed by the well-known Norrish Type I and II reactions, e.g.:

$$\sim\mathrm{C{-}C{=}C{-}\underset{\|}{\overset{}{C}}{-}C}\sim \;\xrightarrow{h\nu}\; \sim\mathrm{C{=}C{-}C{-}\underset{\|}{C}{-}C}\sim$$
$$\text{(C=O in each case)} \qquad \big\downarrow h\nu$$

Norrish Type I and II reactions.

(2) Cross-linking between adjacent α,β-unsaturated carbonyls to produce saturated carbonyls, e.g.:

$$\left.\begin{array}{l}\sim\mathrm{C{-}C{=}C{-}C({=}O)}\sim \\ \sim\mathrm{C{-}C({=}O){-}C{=}C{-}C}\sim\end{array}\right\} \xrightarrow{h\nu} \begin{array}{l}\sim\mathrm{C{-}C{-}C{-}C({=}O)}\sim \\ \quad\;\;\mathrm{|}\quad\;\mathrm{|} \\ \sim\mathrm{C{-}C({=}O){-}C{-}C{-}C}\sim\end{array}$$

During photooxidation of the polymer therefore, the α,β-unsaturated carbonyl groups are converted into saturated ketonic/aldehydic carbonyl groups which are themselves subsequently converted during the photooxidation process to non-luminescent products such as carboxylic acids. This could occur by reactions (1) or (2), although the contribution of (2) for a 'fully saturated' aliphatic polyolefin is likely to be very small.

Finally, evidence supporting the participation of α,β-unsaturated carbonyl impurities in the sunlight-induced oxidation of polyolefins is shown in Fig. 6 for polypropylene. This figure shows a correlation between the concentration of α,β-unsaturated carbonyl groups as measured by fluorescence intensity and the rate of photooxidation of six different batches of commercial polypropylene. It is also seen from the figure that different polymer samples with the same fluorescence intensity exhibited similar rates of photooxidation.

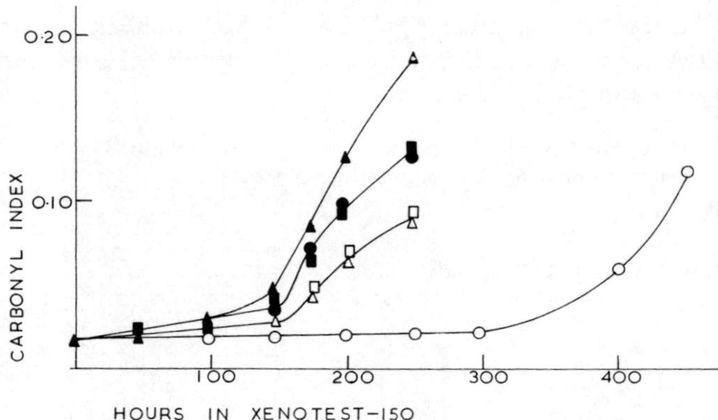

FIG. 6. Variation of carbonyl index versus time of irradiation in a Xenotest-150 weatherometer (50% relative humidity, 45 °C) for different batches of commercial, unstabilised polypropylene films. Before irradiation the samples had relative fluorescence intensities at 340 nm of ▲, 1·0; ■, 0·73; ●, 0·73; △, 0·44; □, 0·44 and ○, 0·25 (Ex λ_{max} = 230 and 285 nm). All polymer samples were vacuum pressed at 190 °C and had melt flow index (MFI) and Ti values of 20 and 25 ppm respectively. (Reproduced from *Polymer*, **18** (1977), p. 968, with permission from IPC Business Press Ltd, UK.)

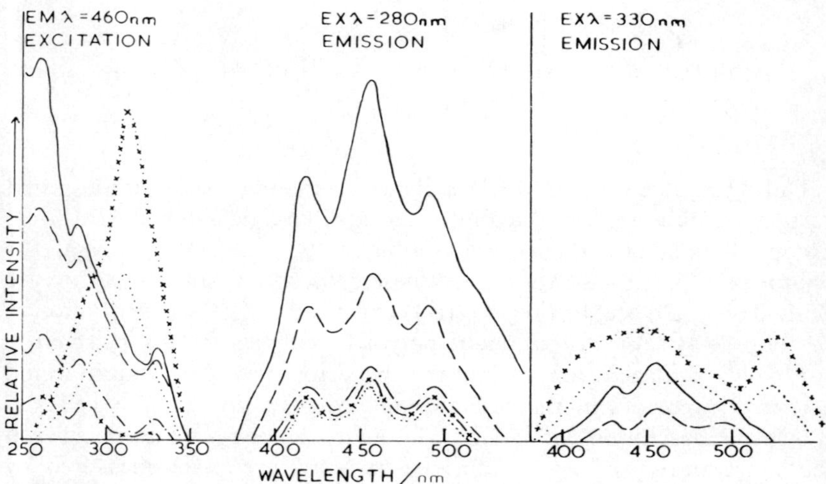

FIG. 7. Changes in the phosphorescence from polypropylene powder after (——) 0 min, (– – –) 15 min, (– · – ·) 30 min, (– × – ×) 45 min, (· · · ·) 60 min and (· × · × · ×) 75 min of heating in air at 130 °C. (Reproduced from *J. Appl. Polym. Sci.*, **20** (1976), p. 2553, with permission from John Wiley & Sons, Inc., New York.)

2.1.4. *Effect of Thermal History*

It is well known that prior thermal oxidation of polyolefins markedly reduces their subsequent photostability.[27–31] Hydroperoxide[27–30] and/or carbonyl groups[31] formed during the thermal oxidation process are believed to initiate photooxidation.

Figure 7 shows the effect of thermal oxidation at 130 °C on the phosphorescence from polypropylene. It is seen that over a period of 45 min there is a gradual reduction in the intensity of the phosphorescence. Further heating however, results in the appearance of a new excitation spectrum with a wavelength maximum at 310 nm due to the formation of saturated ketonic/aldehydic carbonyl groups.[31] Figure 8 shows that on subsequent photooxidation of the thermally oxidised polypropylene samples, a marked reduction in light stability occurs with the appearance of the phosphorescent ketonic/aldehydic carbonyl groups. The fluorescence from

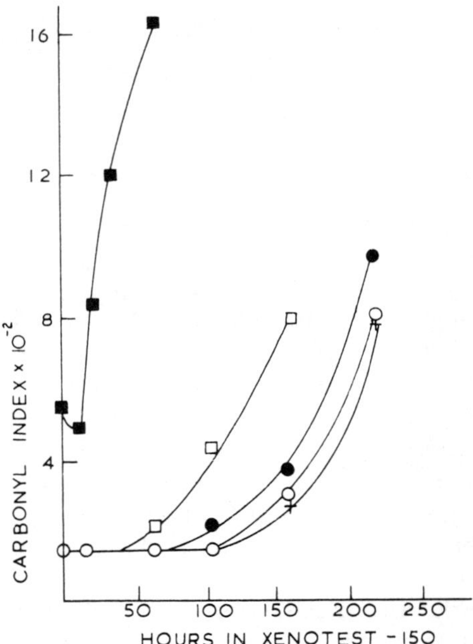

FIG. 8. Carbonyl index versus hours of irradiation in a Xenotest-150 for polypropylene film after (+) 0 min, (○) 15 min, (●) 30 min, (□) 45 min and (■) 60 min of heating in air at 130 °C. (Reproduced from *J. Appl. Polym. Sci.*, **20** (1976), p. 2553, with permission from John Wiley & Sons, Inc., New York.)

the polymer also decreases over the same heating period but shows no new excitation spectrum.[31]

Other workers,[27-30] using infra-red spectroscopy, have found no correlation between carbonyl concentration after processing and the subsequent light stability of polyolefins. However, infra-red spectroscopy measures total carbonyl concentration and therefore can include a variety of carbonyl groups such as esters and carboxylic acids which do not absorb light above 290 nm.[10] Consequently, these groups would not be expected to participate in the sunlight-induced oxidation of the polymer.

2.2. Aliphatic Polyamides (Nylon Polymers)

2.2.1. *Identification of the Luminescent Species*

Simple aliphatic amide compounds with the structures (**VI**) and (**VII**) have been widely used as models for studying the thermal and photochemical oxidation of aliphatic polyamides (nylon polymers).[12-14,32-40]

$$CH_3(CH_2)_3NHCO(CH_2)_4CONH(CH_2)_3CH_3$$

N,N′-di-*n*-butyladipamide

(VI)

$$CH_3(CH_2)_4CONH(CH_2)_6NHCO(CH_2)_4CH_3$$

N,N′-dicaproylhexamethylene-diamine

(VII)

Mild thermal oxidation of both these amide compounds just above their melting points (~180 °C) results in the formation of fluorescent and phosphorescent species whose excitation and emission spectra closely match those of nylon 6,6 polymer[12-14,40] (see Figs. 9 and 10 respectively). Also, the magnitudes of the vibrational splittings of the phosphorescence emissions from the oxidised model amides and the polymer are the same (Table 1). Neither amide shows significant fluorescence or phosphorescence emissions before heating or even after heating under vacuum. The fluorescence and phosphorescence excitation wavelength maxima are different indicating that the species responsible for the emissions are not the same. Also, the fluorescence emission lacks structure and varies with excitation wavelength, indicating that more than one emitting species is involved. Consequently, attempts to identify the species have so far been unsuccessful.[14,50]

The phosphorescence emissions from the oxidised model amides, on the

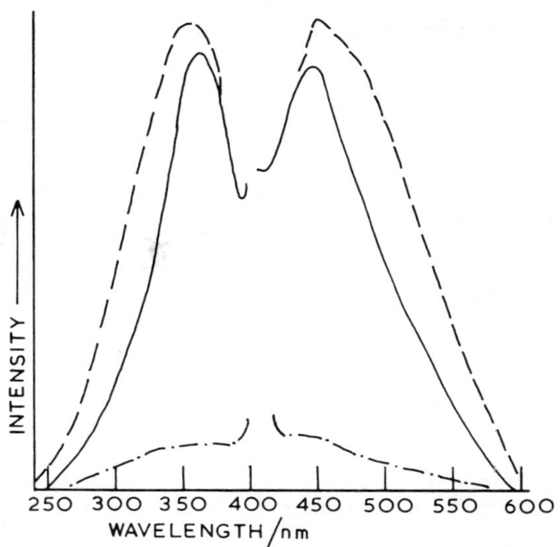

FIG. 9. Comparison of the fluorescence excitation and emission spectra at 300 K and 77 K of (——) nylon 6,6 chip (× 500), (–·–·) model amides before heating and after heating in a vacuum at 180 °C and (– – –) model amides after heating in air at 180 °C (× 1000). (Reproduced from *J. Polym. Sci., Polym. Chem. Ed.*, **12** (1974), p. 2623, with permission from John Wiley & Sons, Inc., New York.)

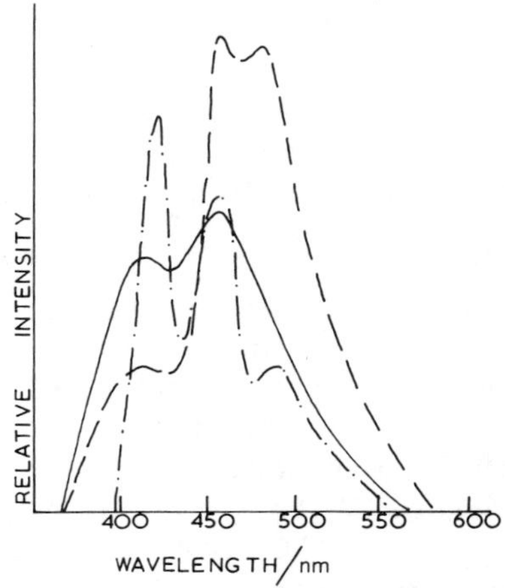

FIG. 10. Comparison of the phosphorescence emission from nylon 6,6 chip (Ex λ = 310 nm) (——) (Sx 300), (Ex λ = 315 nm) (– – –) (Sx 450) with that from the thermally oxidised model amides (Ex λ_{max} = 273 and 350 nm) (–·–·). Sx = sample signal sensitivity. (Reproduced from *J. Photochemistry*, **6** (1976), p. 337, with permission from Elsevier Sequoa, UK.)

other hand, are structured and do not vary with excitation wavelength indicating that only one type of emitting species is involved.[14,40] Further, the magnitude of the vibrational splitting of the emission is the same as that for the $n\pi^*$ emission of a carbonyl group (Table 1).[41] The fact that oxygen is required to produce the phosphorescent species on heating the model amides is consistent with this assignment. Figure 11 shows that the phosphorescence excitation spectrum of one of the thermally oxidised

TABLE 1

WAVENUMBERS OF PEAKS IN THE PHOSPHORESCENCE EMISSION SPECTRA OF NYLON 6,6 CHIP, THERMALLY OXIDISED MODEL AMIDES AND A TYPICAL DIENONE

	λ (nm)	$\nu \times 10^2$ (cm^{-1})	$\Delta\nu \times 10^2$ (cm^{-1})
Nylon 6,6 chip	410–420	244–238	
	450–460	222–217	17·00
	475–480	210–208	
Model amides[a]	417	239·8	
	448·5	222·9	16·85 ± 10
	485	206·2	
Dienone[b]	415·2	240·9	
	446·4	224	16·70 ± 50
	480·9	207·9	

[a] In EPA glass at 77 K.
[b] 4,4′-diphenylcyclohexadienone in EPA glass at 77 K.

model amides matches closely the absorption spectrum of a typical dienone (cf. polyolefin work above). The magnitude of the vibrational splitting of the emission spectrum is also similar to that of a cyclic dienone (Table 1).

It has been suggested that the phosphorescence emission from nylon polymers originates from the amide chromophore itself, with an absorption process for light above 290 nm involving a weak spin-forbidden singlet–triplet transition.[42,43] The evidence for this is based on the observation of weak emission from several 'pure' model amide compounds, but other workers[12–14,40,44–46] have shown that simple 'pure' amide compounds do not phosphoresce. For example, studies by McGlynn and co-workers[45,46] have shown that only structurally perturbed amide chromophores such as **(VIII)** can phosphoresce.

(VIII)

It is difficult to see how structures of this type can be present in the commercial polymer even allowing for the undoubted influence of strong intermolecular hydrogen-bonding forces that exist between adjacent polymer chains.

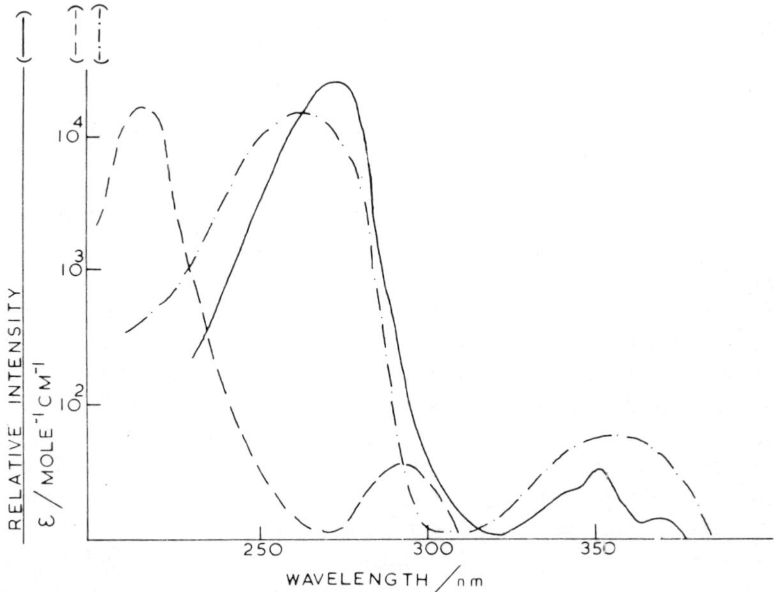

FIG. 11. Comparison of the phosphorescence excitation spectrum of thermally oxidised N,N′-di-*n*-butyladipamide in EPA glass at 77 K (——) with the absorption spectra of a typical enone, pent-3-ene-2-one (– – –), and dienone, 2,6-dimethylhept-2,5-diene-4-one (– · – ·) in *n*-hexane. (Reproduced from *J. Polym. Sci., Polym. Chem. Ed.*, **15** (1977), p. 2793, with permission from John Wiley & Sons, Inc., New York.)

TABLE 2
PHOSPHORESCENCE PROPERTIES OF NYLON POLYMERS

Polymer	*Excitation* λ *(nm)*	*Emission* λ *(nm)*	*Mean lifetime* τ *(s)*[a]
Nylon 6,6 chip	295 max	410 max	2·1
	310	415, 460	1·6, 0·6
	315	420, 460, 480	0·7, 0·25, 0·20
Nylon 6,6 fibre	295 max	450	1·2
	300	460	0·7
	315	470	0·30
Nylon 6 chip	282 max	390(s),[b] 420, 455(s)	1·7, 1·6, 1·1
	300	460	0·80
	310	470	0·58
Nylon 11 chip	269(s), 273	423, 450(s)	1·0, 0·88
	300	460	0·6
	310	465	0·6
Nylon 12 chip	268, 286(s)	363(s), 410	1·0
	300	465	0·6
	310	475	0·6

[a] τ = Time for phosphorescence to decay to 1/e of its initial intensity.
[b] (s) = shoulder.

Finally, the phosphorescence emissions from nylon polymers are long-lived (Table 2) whereas that of the model cyclic dienone (see Table 1) is short-lived (5 ms). This clearly shows the influence of a rigid polymeric environment on the excited α,β-unsaturated carbonyl chromophore as found earlier for the polyolefins. A further point of interest is that the lowest triplet $n\pi^*$ and $\pi\pi^*$ excited states of α,β-unsaturated carbonyl groups are almost degenerate.[22–26] The relative positions of these energy levels are markedly influenced by both structural and environmental factors. Thus, in a strongly hydrogen-bonded matrix such as nylon 6,6 the triplet $n\pi^*$ and $\pi\pi^*$ energy levels would undergo corresponding hypsochromic and bathochromic shifts respectively resulting in an increase in the energy of the $n\pi^*$ state relative to that of the $\pi\pi^*$ state. Consequently, long-lived emission from these dienone chromophores in the molecular backbone of the polymer would be expected. In fact Table 2 shows that for a range of nylon polymers there is an increase in their emission lifetimes with an increase in the extent of their crystallity which is in the order nylon 6,6 > nylon 6 > nylon 11 and 12.

2.2.2. *Effect of Thermal and Photooxidation on the Luminescent Species*
Figure 12 compares the effects of thermal and photochemical oxidation on the phosphorescence from nylon 6,6 polymer in 'chip' form.[14] Thermal oxidation at 180 °C results in a decrease in the intensity of the phosphorescence and the formation of new, longer wavelength excitation and emission spectra. The latter spectra are due to the formation of

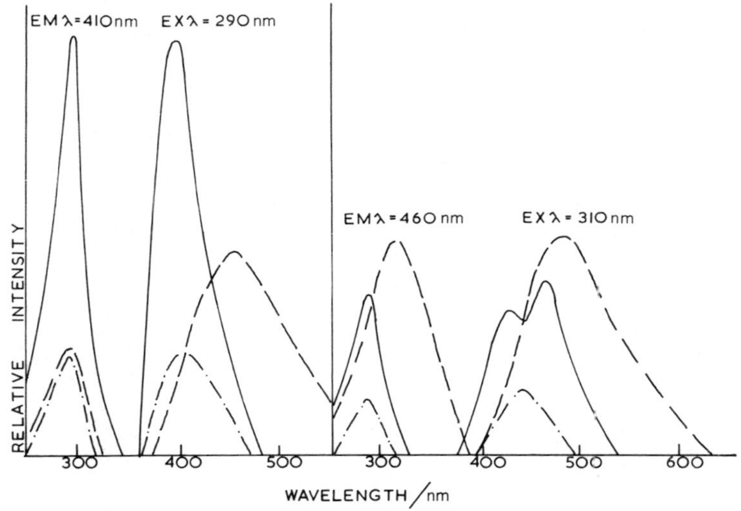

FIG. 12. Phosphorescence spectrum of nylon 6,6 film before (——) and after (– – –) thermal oxidation at 180 °C for 1 h and photooxidation (– · – ·) in a Xenotest-150 weatherometer for 450 h. (Reproduced from *J. Photochemistry*, **6** (1976), p. 337, with permission from Elsevier Sequoa, UK.)

saturated ketonic/aldehydic carbonyl groups.[14] Interestingly, nylon 6,6 fibre, having undergone some degree of thermal oxidation during spinning, exhibits similar longer wavelength phosphorescence emission bands to those of thermally oxidised nylon 6,6 chip (see Table 2).

Photooxidation under sunlight-simulated conditions on the other hand results in an overall decrease in the intensity of the phosphorescence together with a shift of the emission wavelength maximum to shorter wavelengths. This suggests that if the longer wavelength phosphorescing species are formed during photooxidation they would be consumed as intermediates during the overall reaction. Their generation during thermal oxidation would, of course, render the polymer potentially more photoactive by virtue of increasing its capacity for absorbing sunlight in the ultra-violet region 300–350 nm. In fact, Table 3 shows that there is a consumption of these longer wavelength phosphorescing species on photooxidation of prior thermally oxidised nylon 6,6 film.[14] This is shown by a decrease in the relative intensities of the longer wavelength emission bands relative to the emission intensity of the unirradiated film at 455 nm.

TABLE 3
EFFECT OF IRRADIATION ON THE PHOSPHORESCENT OXIDATION PRODUCTS OF NYLON 6,6 FILM (125 μ)

Irradiation time (h)	*Phosphorescence excitation*	*emission λ/nm*	*Mean lifetime s*	*Ratio excitation band intensities (reference Em 455 nm)*
0	280	455	0·82	1·00
	300	465	0·37	0·95
	320	480	0·21	0·88
	360	500	0·17	0·35
25	280	435	1·20	0·85
	300	460	0·40	0·60
	320	475	0·36	0·38
	360	500	0·18	0·08
68	280	430	1·30	0·80
	300	455	0·45	0·53
	320	470	0·30	0·32
	360	500	0·15	0·05
100	280	430	1·30	0·75
	300	450	0·40	0·50
	320	470	0·27	0·31
	360	500	0·15	0·04

It has been reported that photooxidation of polyamides and model amide compounds favours the formation of aliphatic amines and carboxylic acids,[14,37–39] whereas thermal oxidation favours the formation of ketonic/aldehydic carbonyl groups.[14,34–36] The fact that aliphatic amines and carboxylic acids do not phosphoresce[41] accounts for the absence of longer wavelength phosphorescing products during photooxidation. Indeed, Marek and Lerch[47] found that diketones and dialdehydes are reactive intermediates in the photooxidation of nylon 6,6 polymer. Succinic acid was identified as one of the final products. On photooxidation therefore, the dienone impurities undergo reaction (1), discussed above for the polyolefins, to give saturated ketonic/aldehydic

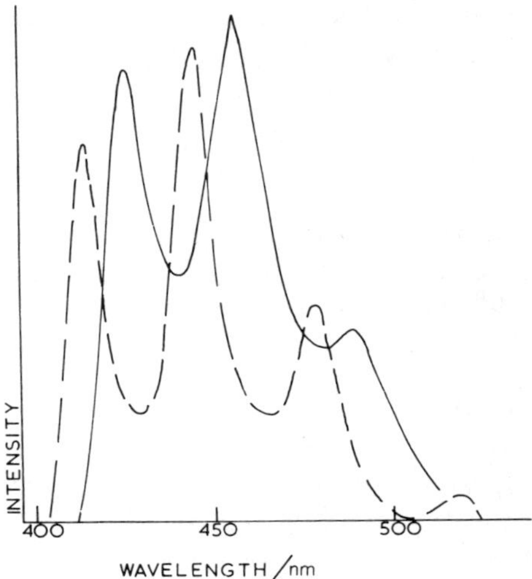

FIG. 13. Phosphorescence emission spectra of an MDI-based polyurethane film (——) and benzophenone (- - -) in 2-propanol (77 K). Excitation $\lambda = 325$ nm. (Reproduced from *J. Appl. Polym. Sci.*, **20** (1976), p. 1441, with permission from John Wiley & Sons, Inc., New York.)

carbonyl groups. These in turn, are photolysed further to give non-phosphorescent dicarboxylic acids.

2.3. Polyurethanes

Polyurethanes synthesised with aromatic diisocyanates are more unstable to light than those synthesised with aliphatic diisocyanates.[48] In particular, commercial polyurethanes based on diphenylmethane 4,4′-diisocyanate (MDI) are the most unstable.

2.3.1. *Identification of Luminescent Species*

Figure 13 shows that, apart from a distinctive 'red-shift', the phosphorescence emission spectrum of an MDI-based polyurethane of the general structure (**IX**) is very similar to that of benzophenone.[49] Also, the splitting of the vibrational structure has a value of 1621 cm^{-1} which is close to that of benzophenone (1612 cm^{-1})

$$\left(\sim ROCO \cdot HN - C_6H_4 - CH_2 - C_6H_4 - NHCO \cdot O \sim \right)_n \qquad \textbf{(IX)}$$

where R = aliphatic polyester or glycol residue.

The presence of a benzophenone type chromophore with the structure (**X**) can be readily envisaged in the molecular backbone of the polymer.

$$\sim\text{N}(\text{H})-\text{C}_6\text{H}_4-\text{C}(=\text{O})-\text{C}_6\text{H}_4-\text{N}(\text{H})\sim \qquad \textbf{(X)}$$

The presence of the substituents in the *para* positions of the benzene ring accounts for the 'red shift' in the phosphorescence emission from that of benzophenone. The polymer also exhibits fluorescence although its identity is unknown.

2.3.2. *Photooxidation and Laser Flash Photolysis*

During irradiation under sunlight-simulated conditions both the fluorescent and phosphorescent species are consumed. Using laser flash photolysis it was possible to observe directly, the primary photochemical processes involving the luminescent species.[49] Figure 14 shows that two transient species are observed on laser flash photolysis of an MDI-based polyurethane film with wavelength maxima at 375 and 580 nm respectively. It is also seen from the figure that the spectrum of the latter transient is

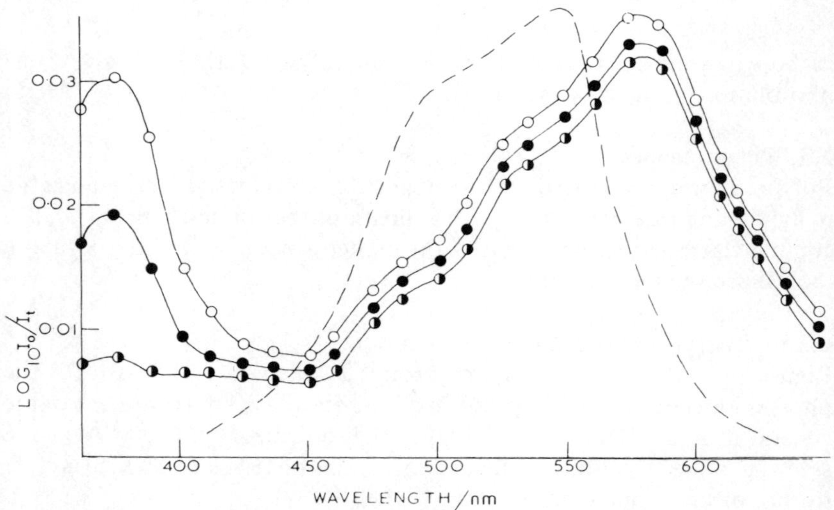

FIG. 14. End of pulse transient absorption observed on laser flash photolysis of an MDI-based polyurethane film (200 μ thickness): (○), after one flash; (●), after two flashes; (◑), after three flashes; (– – –), absorption spectrum of the benzophenone ketyl radical in alcoholic solution. (Reproduced from *J. Appl. Polym. Sci.*, **20** (1976), p. 1441, with permission from John Wiley & Sons, Inc., New York.)

similar to that of the ketyl radical observed from the flash photolysis of benzophenone in alcoholic solution.[50] On repetitive flashing, the former transient at 375 nm gradually disappears while the transient at 580 nm is unaffected.

The observation of the ketyl radical on laser flash photolysis of the MDI-based polyurethane, indicates that the aromatic carbonyl impurity in its photoactive triplet state abstracts a hydrogen atom from the surrounding polymer matrix by reaction scheme (3):

$$\text{\textasciitilde N(H)–C}_6\text{H}_4\text{–C(=O)–C}_6\text{H}_4\text{–N(H)\textasciitilde} \xrightarrow{h\nu} \text{\textasciitilde N(H)–C}_6\text{H}_4\text{–C(=O*)–C}_6\text{H}_4\text{–N(H)\textasciitilde}$$

$$\text{\textasciitilde N(N)–C}_6\text{H}_4\text{–C(=O*)–C}_6\text{H}_4\text{–N(H)\textasciitilde} + \text{\textasciitilde O–C(=O)–N(H)–C}_6\text{H}_4\text{–CH}_2\text{–C}_6\text{H}_4\text{–N(H)–C(=O)–O\textasciitilde}$$

$$\downarrow$$

$$\text{\textasciitilde N(H)–C}_6\text{H}_4\text{–}\dot{\text{C}}\text{(O–H)–C}_6\text{H}_4\text{–N(H)\textasciitilde} \qquad (3)$$

Ketyl radical

$\downarrow [O_2]$

Oxidation

$$+ \begin{cases} \text{\textasciitilde O–C(=O)–N(H)–C}_6\text{H}_4\text{–}\dot{\text{C}}\text{H–C}_6\text{H}_4\text{–N(H)–C(=O)–O\textasciitilde} \\ \text{or} \\ \text{\textasciitilde O–C(=O)–N(H)–C}_6\text{H}_4\text{–CH}_2\text{–C}_6\text{H}_4\text{–}\dot{\text{N}}\text{–C(=O)–O\textasciitilde} \end{cases}$$

The ketyl radical may undergo further reactions resulting in degradation of the polymer. An analogous primary photochemical process is responsible for the benzophenone-sensitised photooxidation of other commercial polymers such as the polyolefins.[51,52]

3. TYPE B POLYMERS

3.1. Poly(ethersulphone)

3.1.1. *Identification of Luminescent Species*

Figures 15 and 16 show that the fluorescence and phosphorescence excitation and emission spectra of a commercial poly(ethersulphone) of the general structure (**XI**) match those of a model diphenyl sulphone.[18]

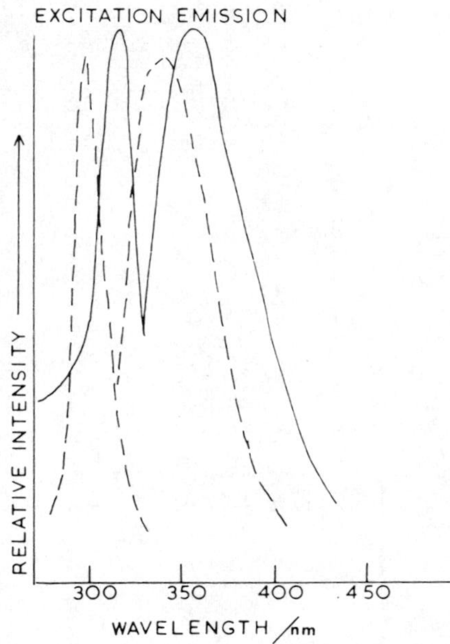

FIG. 15. Fluorescence excitation and emission spectra of poly(ethersulphone) film (——) and 4,4′-dichlorodiphenylsulphone (– – –) in toluene at 300 K. (Reproduced from *J. Appl. Polym. Sci.*, **21** (1977), p. 1129, with permission from John Wiley & Sons, Inc., New York.)

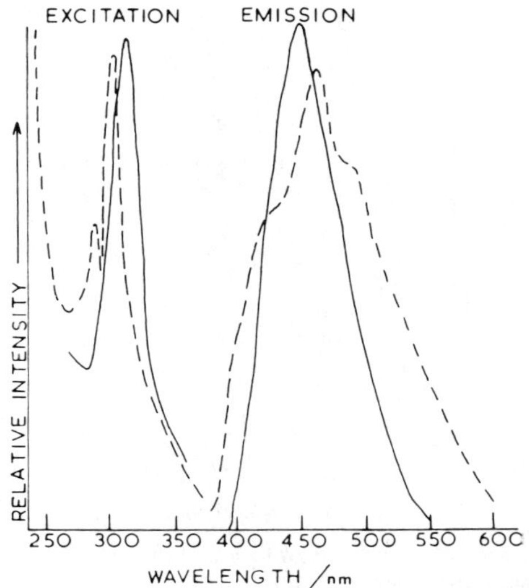

FIG. 16. Phosphorescence excitation and emission spectra of poly(ethersulphone) film (——) and 4,4′-dichlorodiphenylsulphone crystals (– – –) at 77 K. (Reproduced from *J. Appl. Polym. Sci.*, **21** (1977), p. 1129, with permission from John Wiley & Sons, Inc., New York.)

(XI)

Thus, unlike the previous polymers, the species responsible for light absorption and emission is the diphenyl sulphone unit of the polymer (**XII**).

(XII)

3.1.2. *Photooxidation and Laser Flash Photolysis*

Continuous photolysis under sunlight-simulated conditions results in a gradual decrease in the intensity of the fluorescence and phosphorescence emissions from the polymer (Fig. 17).[18] Figure 18 shows that on laser flash photolysis of the poly(ethersulphone) film, transient species are formed

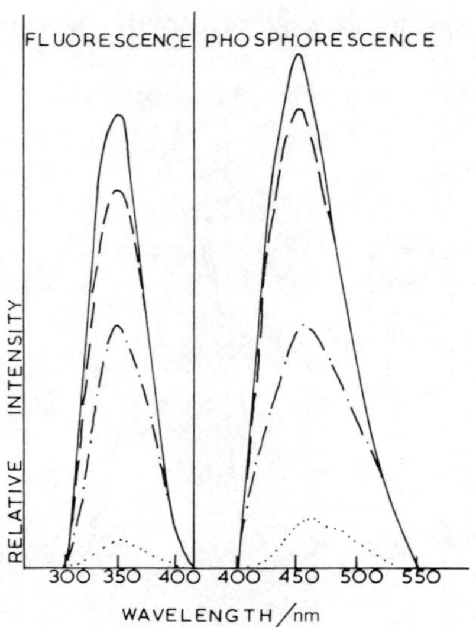

FIG. 17. Decrease in fluorescence and phosphorescence emission intensities from poly(ethersulphone) film after (——) 0 h, (– – –) 52 h, (– · – ·) 100 h and (· · · · ·) 200 h photooxidation in a Xenotest-150. (Reproduced from *J. Appl. Polym. Sci.*, **21** (1977), p. 1129, with permission from John Wiley & Sons, Inc., New York.)

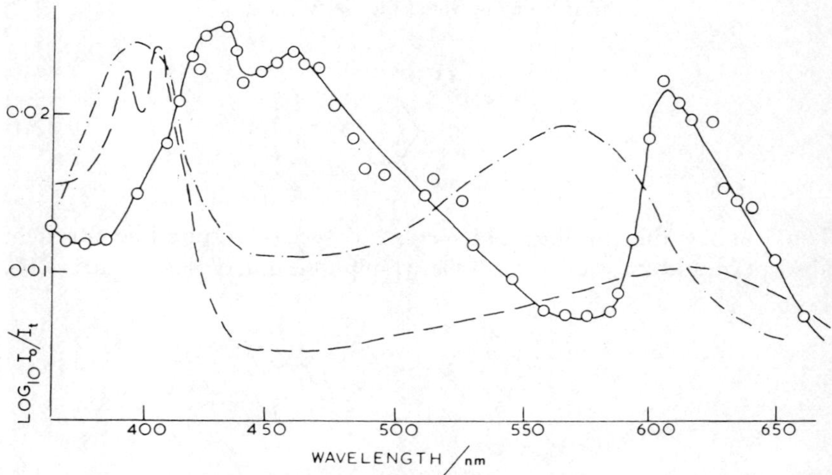

FIG. 18. End of pulse transient absorption observed on laser flash photolysis of poly(ethersulphone) film (100 μ) (○) compared with the absorption spectrum of a phenoxyradical (ϕO·) in liquid paraffin (– – –) and the absorption spectrum of a 2,4,6-triphenoxyradical in 50 % ethanol (– · – ·). (Reproduced from *J. Appl. Polym. Sci.*, **21** (1977), p. 1129, with permission from John Wiley & Sons, Inc., New York.)

with wavelength maxima at 435, 465 and 605 nm. It is also seen from the figure that the spectrum of the transient has the same basic features as those of two simple phenoxyradicals observed from the flash photolysis of phenolic compounds in alcoholic solution.[53] Electron spin resonance studies however have shown that scission can occur either at the phenyl–oxygen bond or at the phenyl–sulphone bond.[54–56] Laser flash photolysis of the model diphenyl sulphone gives no transient species, indicating that sulphoxide radicals are not observable under these conditions.

From laser flash photolysis[18] and ESR data[54–56] therefore, the reaction scheme (4) has been proposed to account for the sunlight-induced degradation of a poly(ethersulphone).

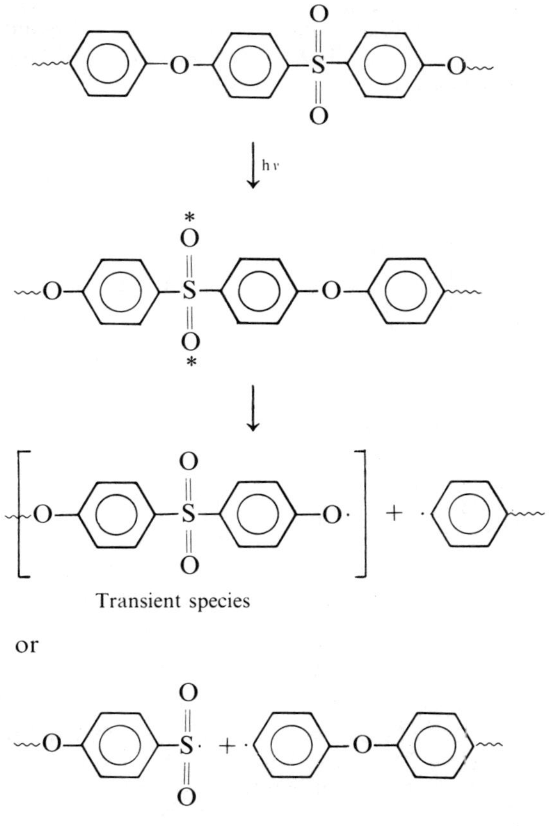

hν

(4)

Transient species

or

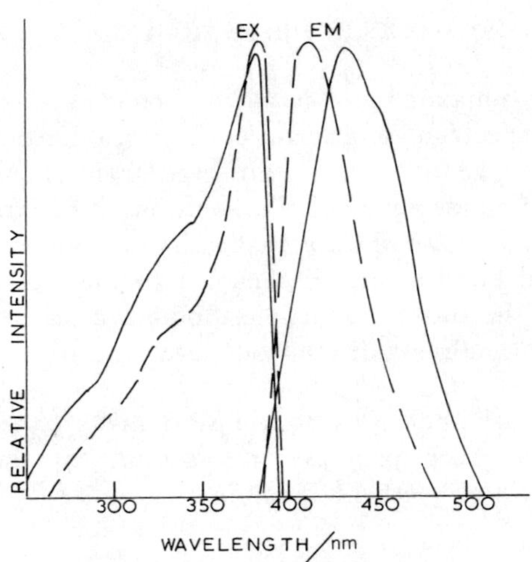

FIG. 19. Comparison of the fluorescence excitation and emission spectra of 300 K of poly(ethylene-2,6-naphthalate) film (——) with that of 2,6-dimethylnaphthalate (– – –) (sample signal sensitivity Sx 0·1). (Reproduced from *J. Appl. Polym. Sci.*, **22** (1978), p. 2085, with permission from John Wiley & Sons, Inc., New York.)

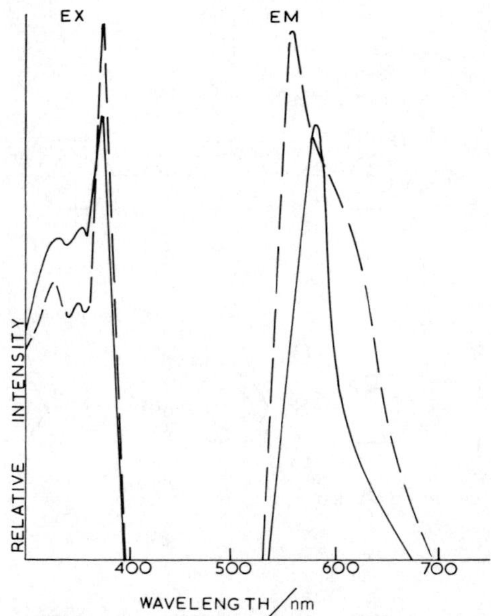

FIG. 20. Comparison of the phosphorescence excitation and emission spectra at 77 K of poly(ethylene-2,6-naphthalate) film (——) with that of 2,6-dimethylnaphthalate (– – –) (sample signal sensitivity Sx 100). (Reproduced from *J. Appl. Polym. Sci.*, **22** (1978), p. 2085, with permission from John Wiley & Sons, Inc., New York.)

3.2. Poly(ethylene-2,6-naphthalate)

3.2.1. *Identification of the Luminescent Species*

Figures 19 and 20 show that the fluorescence and phosphorescence excitation and emission spectra of commercial poly(ethylene-2,6-naphthalate) of the general structure (**XIII**) match those of a model 2,6-dimethylnaphthalate.[17]

$$\left[\sim OC(=O)-C_{10}H_6-C(=O)-O-CH_2-CH_2 \sim \right]_n \qquad \textbf{(XIII)}$$

Thus, the species responsible for light absorption and emission is the unit structure of the polymer backbone.

3.2.2. *Photooxidation and Flash Photolysis*

Continuous photolysis under sunlight-simulated conditions results in a gradual decrease in the intensity of the fluorescence and phosphorescence emissions from the polymer (Fig. 21).[17] Unlike the polyurethane and

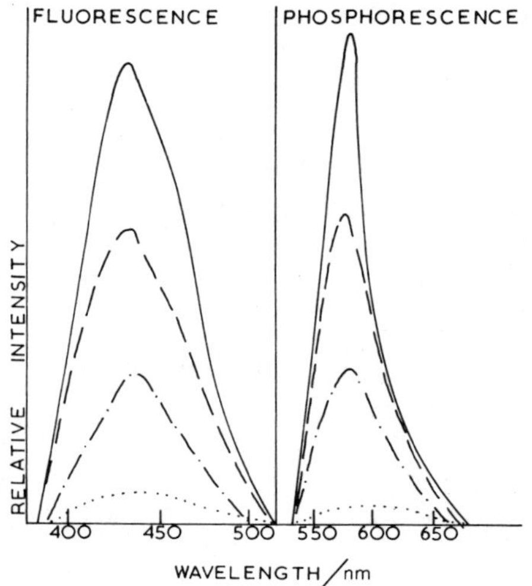

FIG. 21. Fluorescence (300 K) and phosphorescence (77 K) emission spectra of poly(ethylene-2,6-naphthalate) film (——) before and after (– – –) 50 h, (– · –) 100 h and (· · · · ·) 200 h irradiation in a Xenotest-150 weatherometer (Ex $\lambda = 375$ nm). (Reproduced from *J. Appl. Polym. Sci.*, **22** (1978), p. 2085, with permission from John Wiley & Sons, Inc., New York.)

poly(ethersulphone), strong absorption by the poly(ethylene-2,6-naphthalate) in the near ultra-violet region of 300–400 nm prevents the detection of transient species on laser flash photolysis.[17] However, information on the primary photochemical processes in the polymer have been obtained from a flash photolysis study of the model 2,6-dimethylnaphthalate.[17] Figure 22 shows an absorption spectrum of the transient species on flash photolysis of 2,6-dimethylnaphthalate in 2-propanol. From earlier flash photolysis studies[57–59] of aromatic carboxylic acids and esters, the transient species is likely to be a naphthyl radical formed by bond scission at the carbon atom adjacent to the aromatic ring:

$$CH_3{-}O{-}\underset{\underset{O}{\|}}{C}{-}C_{10}H_6{-}\underset{\underset{O}{\|}}{C}{-}O{-}CH_3 \xrightarrow{h\nu} CH_3O{-}\underset{\underset{O}{\|}}{C}{-}C_{10}H_6^{\cdot} + \dot{\underset{\underset{O}{\|}}{C}}{-}OCH_3 \quad (5)$$

Further, the radical decays by a first-order process (see Fig. 22) in 2-propanol, indicating that it must decay by a mechanism of hydrogen atom abstraction from the solvent.[17]

$$CH_3{-}O{-}\underset{\underset{O}{\|}}{C}{-}C_{10}H_6^{\cdot} + (CH_3)_2CHOH \rightarrow CH_3{-}O{-}\underset{\underset{O}{\|}}{C}{-}C_{10}H_7 + (CH_3)_2\dot{C}OH \quad (6)$$

On continuous photolysis, naphthalene is formed by the analogous processes:[17]

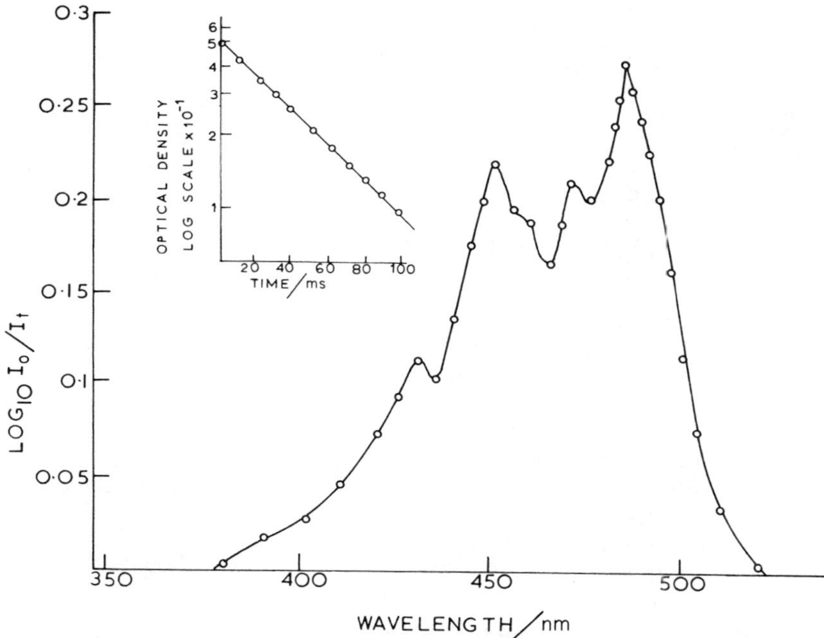

FIG. 22. Transient absorption spectrum produced in the flash photolysis of nitrogen saturated (< 5 ppm oxygen) 10^{-5} M solution of 2,6-dimethylnaphthalate in 2-propanol. The insert shows that a plot of the decay of the transient exhibits first-order kinetics. (Reproduced from *J. Appl. Polym. Sci.*, **22** (1978), p. 2085, with permission from John Wiley & Sons, Inc., New York.)

$$\mathrm{CH_3{-}O{-}C({=}O){-}C_{10}H_7} \xrightarrow{h\nu} \cdot\mathrm{C_{10}H_7} + \cdot\mathrm{C({=}O){-}O{-}CH_3} \qquad (7)$$

$$\cdot\mathrm{C_{10}H_7} + \mathrm{(CH_3)_2CHOH} \rightarrow \mathrm{C_{10}H_8} + \mathrm{(CH_3)_2\dot{C}OH} \qquad (8)$$

This is shown in Fig. 23 in which it is seen that repetitive flash photolysis of 2,6-dimethylnaphthalate results in a product with a fluorescence excitation spectrum closely matching that of naphthalene.[17] Thus, it is evident from

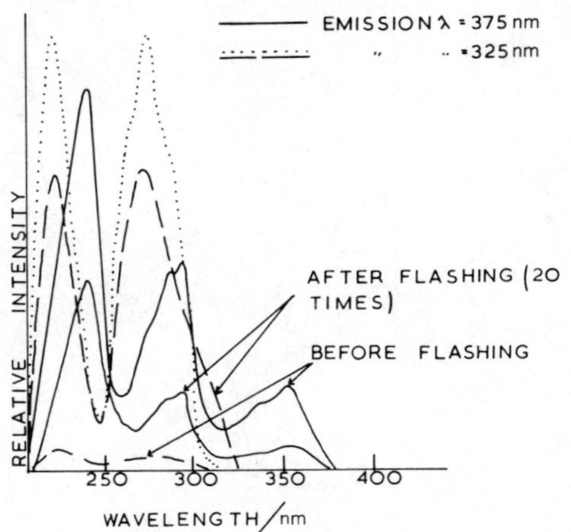

FIG. 23. Comparison of the fluorescence excitation spectrum of (——) 2,6-dimethylnaphthalate (10^{-5} M) (Sx 0·3) and (– – –) its photoproduct before and after twenty photoflashes in the absence of oxygen with that of (· · · · ·) naphthalene (10^{-5} M) (Sx 30) in propan-2-ol. (Reproduced from *J. Appl. Polym. Sci.*, **22** (1978), p. 2085, with permission from John Wiley & Sons, Inc., New York.)

the model-system work that one of the major primary photochemical processes involved in the photodegradation of the polymer is scission of the naphthyl–carbon bond in the unit structure of the polymer:

$$\sim\!O{-}\underset{\displaystyle O}{\underset{\|}{C}}{-}C_{10}H_6{-}\overset{\displaystyle O}{\overset{\|}{C}}{-}O{-}CH_2CH_2\!\sim \xrightarrow{h\nu} \sim\!O{-}\underset{\displaystyle O}{\underset{\|}{C}}{-}C_{10}H_6{\cdot} + {\cdot}\underset{\displaystyle O}{\underset{\|}{C}}{-}O{-}(CH_2)_2\!\sim \qquad (9)$$

REFERENCES

1. CHARLESBY, A. and PARTRIDGE, R. H., *Proc. Roy. Soc.*, **A283** (1965), pp. 312, 329.
2. CHARLESBY, A. and BOUSTEAD, I., *Eur. Poly. J.*, **3** (1967), p. 459.

3. SOMERSALL, A. C. and GUILLET, J. E., *J. Macromol. Sci., Rev. Macromol. Chem.*, **C(B)** (1975), p. 135.
4. ALLEN, N. S., CUNDALL, R. B., JONES, M. W. and MCKELLAR, J. F., *Chem. Ind.*, (*London*) (1976), p. 110.
5. ALLEN, N. S., HOMER, J. and MCKELLAR, J. F., *Chem. Ind.*, (*London*) (1976), p. 692.
6. ALLEN, N. S., HOMER, J. and MCKELLAR, J. F., *J. Appl. Polym. Sci.*, **21** (1977), p. 2261.
7. ALLEN, N. S., HOMER, J., MCKELLAR, J. F. and WOOD, D. G. M., *J. Appl. Polym. Sci.*, **21** (1977), p. 3147.
8. ALLEN, N. S. and MCKELLAR, J. F., *J. Appl. Polym. Sci.*, **22** (1978), p. 625.
9. ALLEN, N. S., MCKELLAR, J. F. and PHILLIPS, G. O., *Am. Chem. Soc., Polym. Preprints*, **18** (1977), p. 375.
10. ALLEN, N. S. and MCKELLAR, J. F., *Polymer*, **18** (1977), p. 968.
11. ALLEN, N. S. and MCKELLAR, J. F., *Brit. Polym. J.*, **9** (1977), p. 302.
12. ALLEN, N. S., MCKELLAR, J. F. and PHILLIPS, G. O., *J. Polym. Sci., Polym. Chem. Ed.*, **13** (1975), p. 2857.
13. ALLEN, N. S., MCKELLAR, J. F. and WILSON, D., *J. Polym. Sci., Polym. Chem. Ed.*, **15** (1977), p. 2793.
14. ALLEN, N. S., MCKELLAR, J. F. and WILSON, D., *J. Photochemistry*, **6** (1976), p. 337.
15. DAY, M. and WILES, D. M., *J. App. Polym. Sci.*, **16** (1972), pp. 175, 191.
16. ALLEN, N. S. and MCKELLAR, J. F., *Makromol. Chem.*, **179** (1978), p. 523.
17. ALLEN, N. S. and MCKELLAR, J. F., *J. Appl. Polym. Sci.*, **22** (1978), p. 2085.
18. ALLEN, N. S. and MCKELLAR, J. F., *J. Appl. Polym. Sci.*, **21** (1977), p. 1129.
19. ABDUL-RADONL, F., CATHERALL, C. L. R., HARGREAVES, J. S., MELLOR, J. M. and PHILLIPS, D., *Eur. Poly. J.*, **13** (1977), p. 1019.
20. PIVOVAROV, A. P., GAK, YU. V. and LUKOVNIKOV, A. F., *Vysokomol. Soedin.*, **A13** (1971), p. 2110.
21. CARLSSON, D. J. and WILES, D. M., *J. Polym. Sci., Poly. Lett. Ed.*, **11** (1973), p. 759.
22. ZIMMERMAN, H. E., BINKLEY, R. W., MCCULLOUGH, J. J. and ZIMMERMAN, G. A., *J. Am. Chem. Soc.*, **89** (1967), p. 6589.
23. CARGILL, R. L., BUNDY, W. A., POND, O. M., SEARS, A. B., SALTIEL, J. and WINTERLE, J., *Mol. Photochem.*, **3** (1971), p. 123.
24. LOUTFY, R. O. and MORRIS, J. M., *Chem. Phys. Letts.*, **19** (1973), p. 377.
25. BEAVAN, S. W. and PHILLIPS, D., *J. Photochemistry*, **3** (1974), p. 349.
26. BELLIS, D., KEARNS, D. R. and SCHAFFNER, K. S., *Helv. Chim. Acta.*, **52** (1969), p. 971.
27. HUTSON, G. V. and SCOTT, G., *Chem. Ind.*, (*London*) (1972), p. 725.
28. MELLOR, D. C., MOIR, A. B. and SCOTT, G., *Eur. Poly. J.*, **9** (1973), p. 219.
29. CHAKRABORTY, K. B. and SCOTT, G., *Polymer*, **18** (1977), p. 98.
30. CHAKRABORTY, K. B. and SCOTT, G., *Eur. Poly. J.*, **13** (1977), p. 731.
31. ALLEN, N. S., HOMER, J. and MCKELLAR, J. F., *J. Appl. Polym. Sci.*, **20** (1976), p. 2553.
32. GOODMAN, I., *J. Polym. Sci., Polym. Chem. Ed.*, **13** (1954), p. 175.
33. GOODMAN, I., *J. Polym. Sci., Polym. Chem. Ed.*, **17** (1975), p. 587.
34. LOCK, M. V. and SAGAR, B. F., *J. Chem. Soc.*, **B** (1966), p. 690.

35. SAGAR, B. F., *J. Chem. Soc.*, **B** (1967), p. 428.
36. SAGAR, B. F., *J. Chem. Soc.*, **B** (1967), p. 1047.
37. MOORE, R. F., *Polymer*, **4** (1963), p. 493.
38. BURNETT, G. M. and RICHES, K. M., *J. Chem. Soc.*, **B** (1966), p. 1229.
39. SHARKEY, W. H. and MOCHEL, W. E., *J. Am. Chem. Soc.*, **81** (1959), p. 3000.
40. ALLEN, N. S., McKELLAR, J. F. and PHILLIPS, G. O., *J. Polym. Sci., Polym. Chem. Ed.*, **12** (1974), p. 2623.
41. CALVERT, J. G. and PITTS, J. N., *Photochemistry*, New York, John Wiley & Sons Inc. 1966.
42. KOENIG, H. S. and ROBERTS, C. W., *J. Appl. Polym. Sci.*, **19** (1975), p. 1847.
43. DELLINGER, J. A. and ROBERTS, C. W., *J. Polym. Sci., Polym. Lett. Ed.*, **14** (1976), p. 167.
44. DEARMAN, H. H., LANG, F. T. and NEELY, W. C., *J. Polym. Sci., Polym. Chem. Ed.*, **7** (1969), p. 497.
45. LARSON, D. B., ARNATT, J. F. and McGLYNN, S. P., *J. Am. Chem. Soc.*, **95** (1973), p. 6928.
46. LARSON, D. B., ARNATT, J. F., SELISKAR, C. J. and McGLYNN, S. P., *J. Am. Chem. Soc.*, **96** (1974), p. 3370.
47. MAREK, B. and LERCH, E., *J. Soc. Dyers Colourists*, **81** (1965), p. 481.
48. SCHOLLENBERGER, C. S. and STEWARD, F. O., *J. Europlastics*, **4** (1974), p. 294.
49. ALLEN, N. S. and McKELLAR, J. F., *J. Appl. Polym. Sci.*, **20** (1976), p. 1441.
50. PORTER, G. and WILKINSON, F., *Trans. Farad. Soc.*, **57** (1961), p. 1686.
51. HARPER, D. J. and McKELLAR, J. F., *J. Appl. Polym. Sci.*, **17** (1973), p. 3503.
52. BURGESS, A. R., *J. Nat. Bur. Studs.*, Circ. 525 (1953), p. 149.
53. LAND, E. J. and PORTER, G., *Trans. Farad. Soc.*, **57** (1961), p. 1885.
54. GESNER, B. D. and KELLEHER, P. G., *J. Appl. Polym. Sci.*, **12** (1968), p. 1199.
55. GESNER, B. D. and KELLEHER, P. G., *J. Appl. Polym. Sci.*, **13** (1969), p. 2183.
56. KHUKHREVA, I. I., DEREVYAGIN, S. V. and BARKOV, A. S., *Vysokomol. Soedin.*, **A14** (1972), p. 1122.
57. JOSCHEK, H. I. and GROSSWEINER, L. I., *J. Am. Chem. Soc.*, **88** (1966), p. 3261.
58. GIVENS, R. S., MATUSZEWSKI, B. and NEYWICK, C. V., *J. Am. Chem. Soc.*, **96** (1974), p. 5547.
59. MITTAL, L. J., MITTAL, J. P. and HAYON, E., *J. Phys. Chem.*, **77** (1973), p. 2267.

Chapter 6

PHOTOCATALYTIC OXIDATION OF POLYOLEFINS

RENÉ ARNAUD and JACQUES LEMAIRE
Université de Clermont-Ferrand, France

SUMMARY

The chemical modifications of a cross-linked polyethylene sample submitted to photocatalysed oxidation are investigated by infra-red spectrophotometry; the spectral changes are compared with those observed in photothermal or thermal oxidation. Zinc oxide and titanium oxides (anatase and rutile forms of the latter) behave similarly as photocatalysts. Initial quantum yields of formation of aliphatic ketones are determined under excitation by monochromatic light; their variations with macroscopic parameters (excitation wavelength, light intensity, percent of the photocatalytic oxide) are studied. Fairly high activation energies are observed in these photocatalysed oxidations. A mechanism of photocatalytic oxidation of a cross-linked polyethylene is proposed; the nature of the photochemical initiation process is discussed and the absence of any inductive effect of the hydroperoxides formed is pointed out. Polyethylene and cross-linked polyethylene behave similarly. The photocatalytic oxidation of isotactic propylene is briefly reported.

1. INTRODUCTION

Photocatalysis by zinc, titanium, magnesium or antimony oxides has attracted attention only recently. This activity, which is related to photoconduction properties, was first studied using small molecular compounds. For example, the influence of zinc or titanium oxides under ultraviolet excitation, on the oxidation of carbon monoxide,[1-3] alkanes,[4–10] olefins,[11] isopropanol[12,13] and furan[14] has been reported. The nature of the intermediate species is still controversial.

In the macromolecular field, it is well known that such white pigments, which are often added to polymeric material, may act as photocatalysts. Photooxidation of polyolefins[15–17] and polyamides[18–23] assisted by titanium or zinc oxides has been reported. Until now, these phenomena have been investigated more or less qualitatively, the main object of the present paper is to present a more quantitative treatment.

Our long-term study has been initiated by a practical problem. The increasing price of carbon black, as well as ecological reasons, have induced a search for insulating materials for aerial cables which do not contain carbon black. Such materials, often cross-linked polyethylenes containing large amounts of white pigments, are highly degradable compared with insulation protected by at least 2% of carbon black. A detailed analysis of the fundamental aspects of the photocatalysed oxidation of cross-linked polyethylene is a necessary prerequisite to the solution of this problem of durability. Research on this topic and its extension to polyethylene and polypropylene are reported.

2. PHOTOCATALYTIC OXIDATION OF CROSS-LINKED POLYETHYLENE

2.1. Analytical Study of the Photothermal and Photocatalysed Oxidation of a Cross-linked Polyethylene Under Polychromatic Excitation

Throughout the whole of this experimental study, polymeric materials have been used in the form of film. The following pigments have been introduced into samples of 'high pressure, low density' polyethylene (PE) (Lupolen 1800H and 2000H, BASF and Lotrene CD 0302 CdF-Chimie): titanium oxide in the anatase form (AT4, Produits Chimiques Thann et Mulhouse, APP2, Tioxide S.A. (0·15 μm)); titanium oxide in the rutile form (RL 90, Produits Chimiques Thann et Mulhouse, RFC 5, Tioxide S.A. (0·17 μm), RMC, Tioxide S.A. (0·23 μm)); Zinc oxide (365·5–Societe Vieille Montagne (0·11 μm)). The figures in brackets indicate the average particle size.

Mixing of the components was carried out in an internal two-rolls mill (Banbury) for 4–5 min at 125–130 °C; the sample was then pressed into a thin film (0·2 mm thick) at 126–128 °C. Cross-linking of the film was achieved by heating with 4% (by weight) of ditert-butyl peroxide at 180 °C for 4 min; the decomposition of this peroxide yields vibrationally excited alkoxy-radicals which react as follows:

$$CH_3{-}C(CH_3)_2{-}O{-}O{-}C(CH_3)_2{-}CH_3 \xrightarrow{\Delta} \left[CH_3{-}C(CH_3)_2{-}O\cdot\right]^*$$

$$\left[CH_3{-}C(CH_3)_2{-}O\cdot\right]^* \xrightarrow[\text{PE}]{60\%} CH_3{-}C(CH_3)_2{-}OH + PE\cdot \rightarrow \text{cross-linking}$$

$$\left[CH_3{-}C(CH_3)_2{-}O\cdot\right]^* \xrightarrow{40\%} CH_3{-}COCH_3 + CH_3\cdot \xrightarrow{\text{PE}} PE\cdot$$

Residual peroxide, t-butanol and acetone diffuse out of the sample as shown by IR spectrophotometry. Neither thermal nor induced oxidation occur during processing or cross-linking (less than 1 carbon atom in 6500 is oxidised). Cross-linking with dicumylperoxide must be avoided in such a study; even in the presence of strongly absorbing pigment, acetophenone, formed in the decomposition of vibrationally excited cumyloxy-radicals, is an efficient photoinducer.

For analytical purposes and for relative kinetic measurements a polychromatic light source may be used if the three following conditions are fulfilled:

First, the rate of photooxidation must not be controlled by diffusion of the oxygen in the sample. In most cases, it is impossible to assess the perturbing effect of oxygen on photolysis mechanisms since oxygen is able to react with almost every intermediate species (excited states, biradicals, monoradicals and even primary photoproducts); through charge-transfer complex formation oxygen may even be able to modify the absorption spectrum of a transparent sample, although this point is controversial. Thus, photolysis experiments in inert gas or in vacuum do not contribute to an understanding of photooxidation, except when a fast, direct photodissociation is occurring.

Second, the polychromatic excitation beam must be filtered to eliminate radiation of wavelengths shorter than a certain limit. This limit may be around 290 nm (natural light) or above, and wavelength effects may be observed by variation of this limit. Most wavelength effects observed can be accounted for by competitive absorption of different additives or by the absorption of primary photoproducts. Under excitation by polychromatic light, vibrational relaxation occurs in any condensed phase, and the photophysics and photochemistry are generally the same as under monochromatic light. Conjugated effects of different wavelengths may only be expected if radicals coming from two different absorbing species are able to interact; such a case has not yet been reported in the literature on polymer photodegradation.

Third, the temperature of irradiated samples must be carefully measured. As shown later, the activation energies of the overall photooxidative processes are high (see Section 2.2) and large variations of the rates with temperature are to be expected. In most conventional exposure systems for testing accelerated photoageing (xenotest, fade-o-meter, weather-o-meter, ...) the temperature of a fast moving sample cannot be easily determined. A new system has been built which allows the use of high intensity light and a strict control of the temperature of the irradiated samples.

The photothermal oxidation of a transparent sample irradiated at 50 °C by wavelengths longer than 305 nm may be followed by IR spectrophotometry. Indeed, for cross-linked polyethylene, this spectroscopic method is the only accurate technique available. A Perkin Elmer model 180 IR spectrophotometer has been used in this study.

Figure 1 shows that in the carbonyl region, aliphatic ketones are initially observed at 1718 cm^{-1}, then acids (1710 cm^{-1}), esters (1735 cm^{-1}) and lactones (1780 cm^{-1}) are formed in secondary processes as well as isolated double bonds (1640, 1410, 990, 915 and 905 cm^{-1}). Similar assignments have previously been proposed.[24,25] In the region of hydroxyl stretching frequencies, a broad absorption band is observed with a maximum around 3380 cm^{-1}. This absorption disappears on heating the sample in vacuum above 120 °C, but it is invariant on vacuum photolysis; such absorption must be attributed to hydrogen-bonded alcohols which are esterified at 120 °C.

If cross-linked polyethylene is thermally oxidised around 127 °C, a broad band appears with a maximum at 3420 cm^{-1} and a sharp absorption can be observed at 3550 cm^{-1}. On heating the oxidised sample above 120 °C under vacuum, the sharp band at 3550 cm^{-1} disappears and the maximum of the broad absorption shifts towards 3400 cm^{-1}; an unassociated hydroxyl vibration is observed at 3635 cm^{-1} and is attributed to monomeric alcohol. The activation energy of disappearance of the 3550 cm^{-1} band is 30 $\pm$ 2 kcal/mol. By irradiation of the pre-oxidised sample under polychromatic light ($\lambda > 305$ nm), absorption at 3550 and 3420 cm^{-1} decreases. Thus, the 3550 cm^{-1} band must be attributed to a monomeric hydroperoxide and the 3420 cm^{-1} band to hydrogen-bonded hydroperoxide. The activation energies of dissociation of such compounds are in the range of 30–40 kcal/mol and photodissociation of hydroperoxide occurs at wavelengths longer than 300 nm.[26] In a transparent polyethylene sample, the stationary

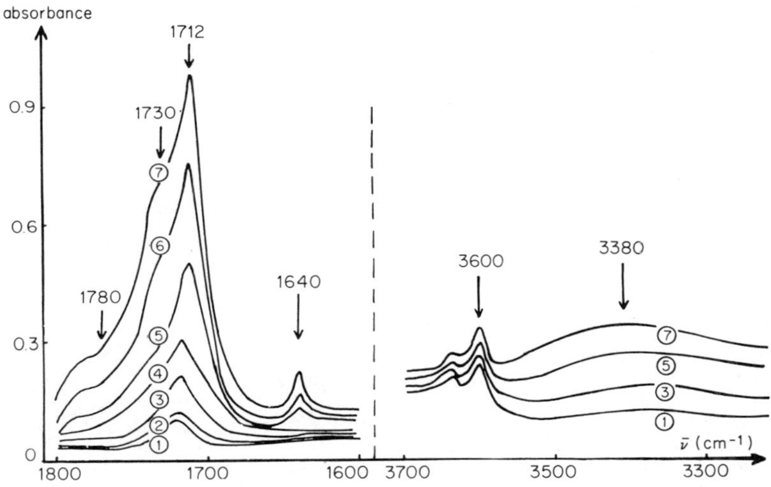

FIG. 1. IR spectra of cross-linked polyethylene samples (0·2 mm) during photothermal oxidation; polychromatic light $\lambda > 305$ nm; temperature of samples, 60 °C.

concentration of hydroperoxide is far higher in thermal oxidation (even around 130 °C) than in photothermal oxidation.

When a light-absorbing pigment such as TiO_2 or ZnO is added to a transparent polyethylene sample, the evolution of the IR spectra during photooxidation is quite different (Fig. 2). In the carbonyl region a sharp band can be observed at 1716–1718 cm^{-1}. Post-photochemical treatment of the sample with an alcoholic solution of NaOH shows that the contribution of acids is of minor importance. No double bonds are revealed by absorption around 1640 cm^{-1}. In the hydroxyl stretching region, absorption of unassociated and hydrogen-bonded hydroperoxides is observed at 3550 and 3420 cm^{-1} respectively. In the photocatalysed oxidation of cross-linked polyethylene, hydroperoxides are the primary products and formation of aliphatic ketones occurs in a secondary step.

These compounds are photochemically protected by the absorbing pigments. Hydroperoxides accumulate and reach a higher stationary concentration than in the transparent sample. Photochemical processes in ketones are also inhibited, specially Norrish Type II processes; no unsaturation appears as a tertiary step. (In a preliminary study it has been shown that the terminal double bonds of polyethylene do not disappear

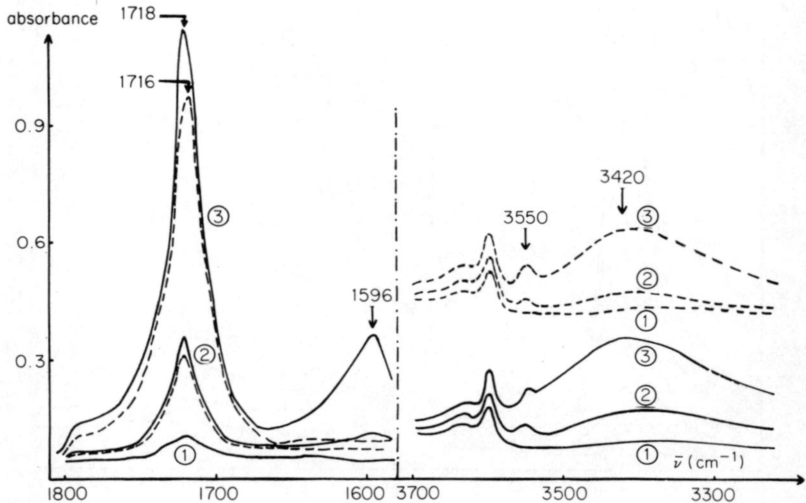

FIG. 2. IR spectra of cross-linked polyethylene samples (0·2 mm) during photocatalytic oxidation; (——), +8% ZnO; (– – –), +8% TiO_2; polychromatic light $\lambda > 305$ nm; temperature of samples, 60 °C.

during the same period of photocatalytic oxidation.) Nevertheless, acids, esters and lactones are the final products even in photocatalytic oxidation since, for example, zinc carboxylate can be observed (at 1596 cm^{-1}) when ZnO is present.

Figure 3 shows the relationship between absorbance at 3420 cm^{-1} and at 1718 cm^{-1} during thermal and photocatalysed oxidation; ΔD represents the variation of absorbance between oxidised and unoxidised films. In each case, the concentration of carbonyl compounds (essentially ketones) remains proportional to the concentration of hydroperoxides. Although hydroperoxides are certainly the primary products, these curves show that ketones initially follow the same law of appearance. In transparent samples, hydroperoxides are photolysed and the lower curve of Fig. 3 represents the contribution of alcohols to the 3420 cm^{-1} absorption.

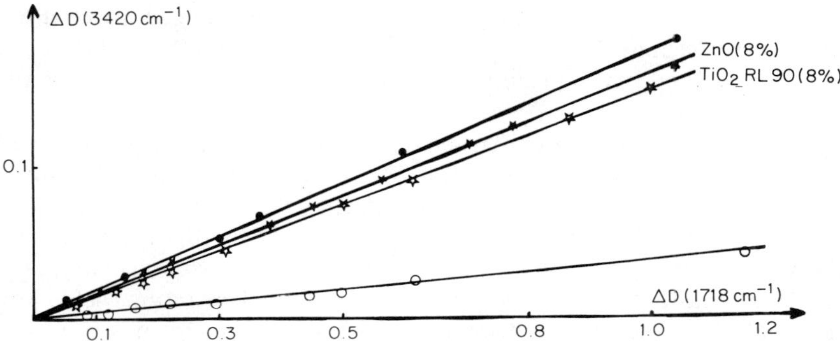

FIG. 3. Relationship between absorbance at 3420 cm^{-1} and at 1718 cm^{-1} during thermal (●), photothermal (○) and photocatalytic (★, ☆) oxidation; polychromatic light $\lambda > 305$ nm; temperature of samples, 60 °C.

2.2. Kinetic Study of the Photocatalytic Oxidation of a Cross-linked Polyethylene Under Monochromatic Light. Determination of Absolute Quantum Yields

Absolute quantum yields of photocatalytic oxidation can be measured under monochromatic irradiation. A 1600 W, high pressure xenon source was fitted with a GM 250 grating monochromator (Schoeffel); the light supplied has a spectral width of 3 or 10 nm. Actinometric measurements of the incident intensity are based upon the photoisomerisation of *trans*-azobenzene[27] or 1-phenyl-azonaphthalene.[28] At wavelengths longer than 400 nm, the conventional ferri-oxalate actinometer has been used.[29–31] Data are presented in Table 1.

TABLE 1

λ nm	I_0 *photons/cm^2 s*	
	Spectral width (3 nm)	*Spectral width* (10 nm)
270	$6{\cdot}2 \times 10^{13}$	—
290	$1{\cdot}46 \times 10^{14}$	—
300	$2{\cdot}8 \times 10^{14}$	$0{\cdot}9 \times 10^{15}$
320	—	$1{\cdot}4 \times 10^{15}$
340	—	$2{\cdot}07 \times 10^{15}$
365	—	$3{\cdot}3 \times 10^{15}$
380	—	$3{\cdot}38 \times 10^{15}$
400	—	$3{\cdot}85 \times 10^{15}$
420	—	$3{\cdot}5 \times 10^{15}$
450	—	$4{\cdot}2 \times 10^{15}$

The determination of quantum yields needs prolonged irradiation (100–150 h). Meanwhile the sample is heated to 50 °C by IR irradiation. It has been checked that no significant oxidation is occurring under this IR excitation alone.

The complexity of the overall mechanism of oxidation of polyethylene to form ketones prevents any calculation of the law of variation of the initial rate of appearance of carbonyl compounds with the absorbed intensity. We have shown experimentally that this initial rate, V_0, is proportional to the absorbed intensity I_a (the straight line representing the variation of $\log_{10} V_0$ with $\log I_a$ has a slope of 0·96). Thus the measured quantum yields are independent of the absorbed intensity. The measurement of significant quantum yields of photocatalytic oxidation defined as the number of ketonic groups formed per photon absorbed, meets two difficulties:

First, when included in a polymeric matrix, a photoactive pigment can absorb or scatter the light. Computation or measurement of the true absorbed intensity are difficult when these two phenomena occur competitively.[32,33] However, relative evaluation of the scattered light can be carried out using an integrating sphere and the range of total absorption can be deduced. For example above 0·6 % TiO_2 (rutile) or 0·6 % ZnO, total absorption by the pigments is observed up to 380 nm. TiO_2 and ZnO are indeed absorbing to 480 nm and 430 nm respectively, but in this range determination of quantum yields is hazardous.

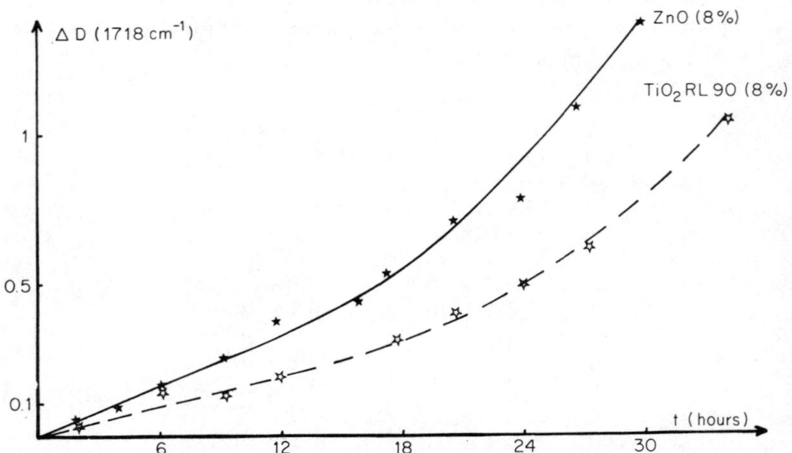

FIG. 4. Absorbance at 1718 cm^{-1} during photocatalytic oxidation; polychromatic light $\lambda > 305$ nm; temperature of samples, 60 °C.

TABLE 2
ABSOLUTE QUANTUM YIELDS OF PHOTOCATALYTIC OXIDATION OF CROSS-LINKED POLYETHYLENE

Additives (percent by weight)	λ_{exc} *(nm)*					
	290	300	320	340	365	380
ZnO 365-5, 8%		0·003	0·0018	0·0012	0·0012	0·0006
ZnO 365-5, 0·6%		0·0012			0·0002	
TiO_2 RL 90, 16%		0·0032		0·0014		0·0010
TiO_2 RL 90, 8%		0·0026	0·001	0·0006	0·0006	0·0002
TiO_2 RL 90, 0·6%		0·0012			0·0002	
TiO_2 AT4, 16%		0·0044		0·0014		0·0012
Carbon black high absorption furnace 0·12% + TiO_2 RL90, 8%	0·0048	0·0052	0·0044	0·0032	0·0018	0·0006

Second, the kinetics of appearance of hydroperoxides and ketones shows self-acceleration (Fig. 4); initial quantum yields must therefore be measured. Conventionally, these quantum yields have been determined for an absorbance at 1718 cm^{-1} equal to 0·5 (1 oxidised carbon atom in 650).

Table 2 summarises the dependence of initial quantum yields of photocatalytic oxidation at 50 °C upon excitation wavelength and with the different additives. The concentration of carbonyl groups is measured from absorbance at 1718 cm^{-1}, assuming a molar extinction coefficient of 250 mol^{-1} . litre . cm^{-1} at 50 °C, a commonly accepted value.[34,35] Due to the self-accelerating character of the reaction, relative errors are estimated to be 10%.

The quantum yield of photothermal oxidation of a transparent sample does not appear in this table since it is not an intrinsic property of the sample, absorption being essentially due to impurities. In any case, the quantum yield of photothermal oxidation is lower than 0·0002.

The values reported in Table 2 prompt the following comments:

In the field of polyolefin photooxidation, no equivalent data have been reported in the literature. These overall quantum yields are rather low. According to F. R. Mayo *et al.*, the average kinetic chain-length for oxidation of polyethylene is less than unity.[36]

As the excitation wavelength increases above 300 nm, the photon efficiency decreases continuously. The same wavelength effect has been reported in the small molecular field for the oxidation of isobutane photocatalysed by TiO_2.[4] Juillet *et al.* have shown by calorimetric measurements that TiO_2 absorbs to 420 nm.[4] Indeed the photocatalytic activity of TiO_2 and ZnO extends to 480 and 436 nm respectively. Figure 5 shows the changes in absorbance at 1718 cm^{-1} during monochromatic excitation with different wavelengths using both oxides. As already mentioned, the coexistence of absorption and scattering prevents any accurate measurement of the quantum yield.

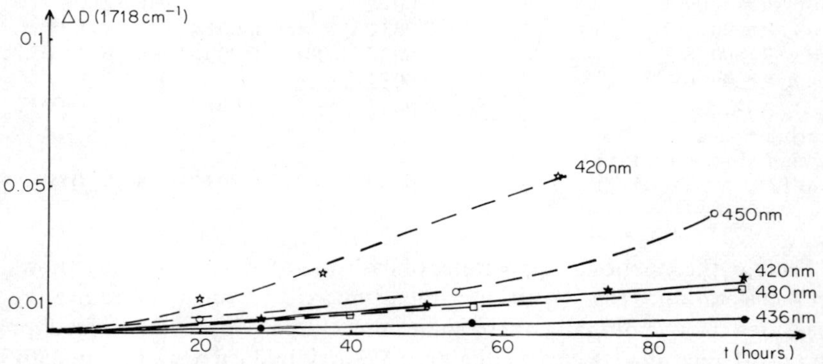

FIG. 5. Long wavelengths: effects in photocatalysed oxidation; (——), +8% ZnO; (— —), +8% TiO_2; temperature of samples, 50 °C.

Another wavelength effect may be pointed out. Although the photocatalytic oxidation promoted by wavelengths shorter than 400 nm is superficial, the photooxidation under visible light ($400 < \lambda < 480$ nm for TiO_2) progresses deeply into a thick sample. Samples (30 mm) were irradiated by two polychromatic lights with wavelengths longer than 305 and 395 nm respectively. After irradiation, a series of 0·2 mm films were cut from the surface of the irradiated samples. The absorbance of these films at 1716 cm^{-1} are shown in Fig. 6 as a function of *e*, distance from the irradiated surface. Oxidation can be observed in both cases and must be attributed to the penetration of visible light ($395 < \lambda < 450$ nm) through scattering. If a similar thick sample is irradiated monochromatically between 300 and 380 nm, similar analysis shows only superficial oxidation ($e < 0{\cdot}2$ mm).

The photocatalytic activities exhibited by TiO_2 anatase and rutile and by ZnO are not very different. In Fig. 7, five different grades of TiO_2 are compared under filtered polychromatic light ($\lambda > 305$ nm). In this experiment, wavelengths between 300 and 400 nm are totally absorbed by

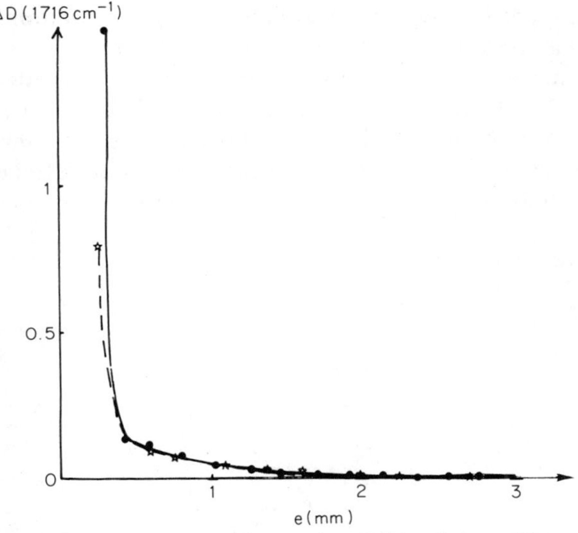

FIG. 6. Deep photocatalysed oxidation in visible light; ΔD, variation of absorbance at $1716\,cm^{-1}$ of 0·2 mm films cut from a 30 mm sample; *e*, distance to the irradiated surface; ●, excitation by wavelengths longer than 305 nm; ☆, excitation by wavelengths longer than 395 nm.

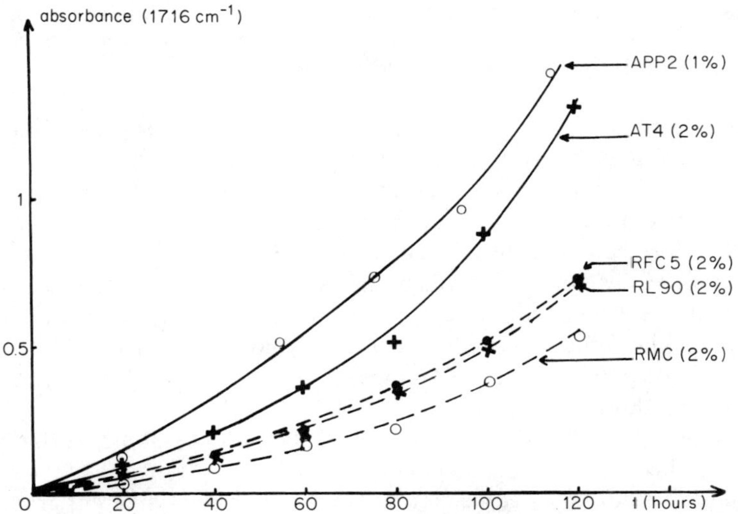

FIG. 7. TiO_2–photocatalytic oxidation of a cross-linked polyethylene; polychromatic light $\lambda > 305$ nm; temperature of sample, 50 °C.

the pigment (2 % by weight). The photocatalysed oxidation is somewhat faster with anatase than with rutile. According to McKellar *et al.*, the anatase form, unlike the rutile form, sensitises the photodegradation of polyolefins. It was suggested that this photosensitising action is linked with the ability of anatase to quench the phosphorescence from polyolefins; rutile is unable to quench this phosphorescence.[15] As a matter of fact, many grades of titanium oxide are available and comparisons between

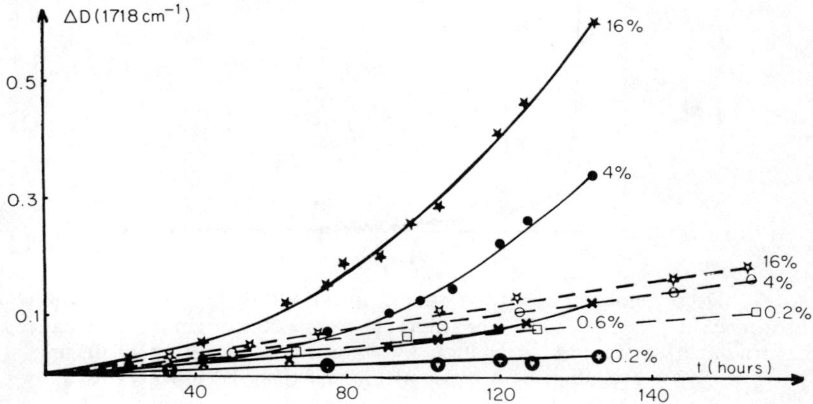

FIG. 8. Pigment concentrations effects in photocatalytic oxidation; (——), ZnO (0·2–16 %); (— —), TiO_2 RL 90 (0·2–16 %); polychromatic light $\lambda > 305$ nm; temperature of sample, 35 °C.

photoactivity of both forms must take into account the surface treatment of the pigment and the average size of the particles.

Above 0·6 % the content of pigment has no influence on the absorption of wavelengths in the range 300–380 nm. The photocatalytic activity should not therefore be concentration dependent. However a small effect of the content of oxides can be observed both in Table 2 and in Fig. 8. In the latter case, samples were irradiated at 35 °C by filtered polychromatic light ($\lambda >$ 305 nm). The influence of ZnO content is greater than for TiO_2. The influence of oxide content is even more important under visible light ($\lambda >$ 395 nm) as shown in Fig. 9.

It is also clear that self-acceleration of the photocatalytic oxidation is more marked at high oxide content.

We have reported a detailed study of the influence of different grades of carbon black on the photothermal oxidation of cross-linked polyethylene.[37] If the temperature of the sample is maintained above 70 °C, introduction of small quantities of carbon black induces a chemical degradation which is at a maximum between 0·3 and 0·7 % for all grades of carbon black studied (Special Furnace Black, High Abrasion Furnace, Intermediate Super Abrasion Furnace, Semi-reinforcing Furnace, Medium Thermic). Conversion of photon energy into thermal energy and delocalisation of this energy in the polymer are the key factors in the degradation mechanism. At low carbon black content (<1 %), localised

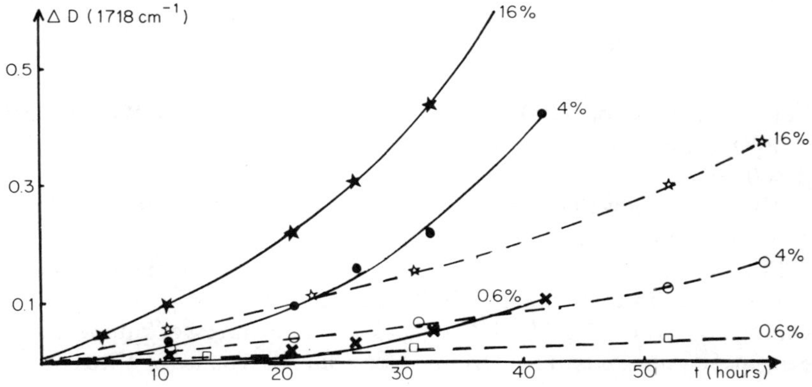

FIG. 9. Pigment concentrations effects in photocatalytic oxidation; (——), ZnO; (– – –), TiO_2 RL 90; polychromatic light $\lambda > 395$ nm; temperature of sample, 60 °C.

heating of the sample occurs and fast thermal oxidation may be observed. At higher contents (between 1 and 10 %) thermal conduction inhibits the oxidation; only stress deformation of the polymer is observed. At contents higher than 10 %, the local temperature of absorption sites is not different from the average temperature of the sample; neither chemical oxidation, nor deformation can be observed.

By simultaneous addition of TiO_2 and a low carbon black content, polyethylene is very photodegradable. No fundamental explanation has been proposed for this observation. However the incompatibility of these two additives must be carefully considered if light protection is needed.

Such a degradable sample can be used as a fast, solid actinometer, especially for comparing different light sources. In most solar light simulators for photoageing tests, a high pressure xenon source is used and the commonly accepted reason is the similarity of the continuum of xenon and the spectrum of the solar light. However the comparison of different sources must be more quantitative. For example, we have determined the average quantum yield for photooxidation of a sample (8 % TiO_2 + 0·12 % carbon black) between 300 and 400 nm for a high pressure xenon source, a very high pressure mercury source and solar light. This average quantum yield is computed from the measured quantum yields ϕ_i reported in Table 2 and from the relative numbers of photons α_i in spectral bands of 20 nm width.

$$\bar{\phi} = \sum_{300}^{400} \alpha_i \phi_i$$

For xenon (Osram XBO 2500 W), $\bar{\phi} = 0{\cdot}0030$; mercury (Osram HBO 500 W), $\bar{\phi} = 0{\cdot}0026$; solar light (noon, summer, Chicago, USA), $\bar{\phi} = 0{\cdot}0018$.

The results show that the simulation of solar light by a xenon source is no better than by a mercury source.

For practical reasons activation energies have been measured under filtered polychromatic light ($\lambda > 305$ nm), the temperature of the sample varying between 27 and 70 °C (Fig. 10). Initial rates of appearance of carbonyl compounds are determined from an absorbance at 1718 cm^{-1} up

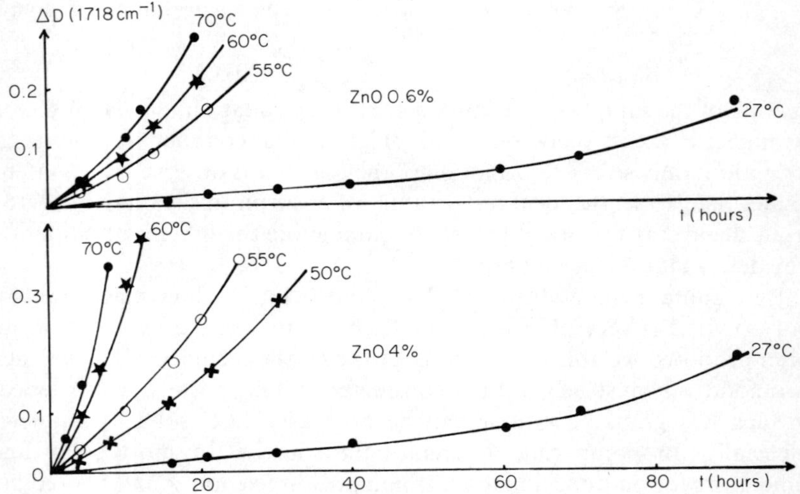

FIG. 10. Temperature effects in a ZnO photocatalysed oxidation; polychromatic light $\lambda > 305$ nm.

to 0·5. The slope of the plots $\log V_0$ versus $1/T$ yields the following activation energies:

$$E = 17{\cdot}5\,\text{kcal/mol for } 16\%\ \text{TiO}_2 \text{ anatase}$$
$$E = 18{\cdot}9\,\text{kcal/mol for } 1\%\ \text{TiO}_2 \text{ anatase}$$
$$E = 18{\cdot}2\,\text{kcal/mol for } 4\%\ \text{ZnO}$$

These activation energies are rather high for an overall reaction induced photochemically.

2.3. Study of the Initiation Step of the Photocatalysed Oxidation of a Cross-linked Polyethylene

The widths of the forbidden bands in TiO_2 and ZnO are 3·5 eV and 3·2 eV respectively, which corresponds to absorption of wavelengths shorter than 350 and 380 nm. Under u.v. irradiation, the absorbing pigments generate electron-positive hole pairs. It is well known that electrons are trapped by chemisorbed molecular oxygen.

$$O_2\,(\text{ads}) + e^- \rightarrow O_2^-\,(\text{ads})$$

The concentration of O_2^- species can be measured by ESR; when the photoactivity of TiO_2 decreases to zero between 110 and 150 °C, the concentration of O_2^- also decreases.[4] F. Juillet *et al.* assumed that simultaneous formation of O_2^- and of a positive hole delocalised in the crystal lattice is necessary for photoactivity of TiO_2, while loss of photoactivity at wavelengths longer than 420 nm is due to the fact that longer wavelengths induce formation of O_2^- and of a hole localised on impurities.[4]

Intervention of the atomic species O^- (ads) has been proposed. The Research Group at the Institut de la Catalyse de Lyon through measurements of thermoelectronic work function[6] and of isotopic exchange[8,9] has assumed the following steps to occur:

$$O_2\,(\text{ads}) + 2e^- \rightarrow 2O^-\,(\text{ads})$$
$$O^-\,(\text{ads}) + \oplus \rightarrow O^*\,(\text{ads})$$
$$O^*\,(\text{ads}) \rightarrow O^*\,(\text{gas})$$

The 'activated' atomic oxygen, chemisorbed or not, is then able to abstract a hydrogen atom and initiate oxidation of isobutane.

According to Bickley and Jayanti, the reactive species would be somewhat different in the photocatalytic oxidation of isopropanol.[12] The following processes have been proposed, in which chemisorbed water (more or less dissociated) plays a significant role.

$$\oplus + OH^- (ads) \rightarrow OH\cdot (ads)$$
$$e^- + O_2 (ads) \rightarrow O_2^- (ads)$$
$$OH\cdot (ads) + O_2^- (ads) \rightarrow HO_2\cdot (ads) + O^- (ads)$$

The reactive species would be $HO_2\cdot$ and O_2^-, which are both able to abstract a hydrogen atom or a proton from isopropanol.

Molecular singlet oxygen has also been proposed as an oxidising species by P. Pappas and R. M. Fischer.[14] This proposal was based upon the fact that oxidation of furane photocatalysed by TiO_2 gives rise to the same intermediate product as singlet oxygen.

$$\xrightarrow[O_2(^1\Delta g)]{h\nu,\ TiO_2} \qquad \xrightarrow{CH_3OH}$$

Even with small molecular compounds, reactive species able to abstract a hydrogen atom in the initiation process of the oxidative mechanism have

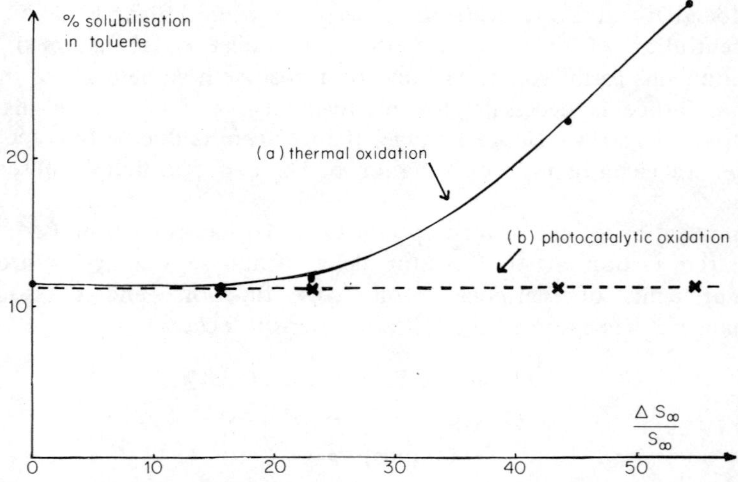

FIG. 11. Dependence of solubilisation in toluene upon $\Delta S_\infty / S_\infty$ (S_∞ = ultimate strength) in the thermal and photocatalysed oxidation of a cross-linked polyethylene sample (8% TiO_2 RL 90 + 0·12% carbon black); (——), thermal oxidation at 150 °C; (- - - - -), photocatalytic oxidation in polychromatic light ($\lambda >$ 305 nm) at 50 °C.

not been definitively identified but radicals like OH·, $HO_2^{\cdot}$, atomic oxygen O·, or singlet oxygen O_2^* (Δ_g) must be considered.

In the macromolecular field, it would be even more difficult to identify the reactive species initially formed. However, a better knowledge of the primary hydrogen abstraction on the polymer chain can be achieved if, during the course of oxidation, the changes in various different physical properties of the sample are followed. We have compared changes in solubility in toluene at 80 °C, Young's modulus and ultimate strength and elongation at break (%), when the same sample of cross-linked polyethylene is submitted to thermal, photothermal and photocatalytic oxidation.

If a cross-linked polyethylene is thermally oxidised, for example at 150 °C, in air, the solubility in toluene increases as ultimate strength and elongation at break decrease (Fig. 11, curve (a)). The Young's modulus decreases. In such an oxidation, an initiation process involving abstraction of tertiary hydrogen atoms is specifically implied since, as represented below, this leads to the links between different chains being destroyed.

$$\sim\!C(H)\!\sim \xrightarrow[\text{abstraction}]{\text{hydrogen}} \sim\!\dot{C}\!\sim \xrightarrow{+O_2} \sim\!C(OO\cdot)\!\sim$$

$$\downarrow$$

$$\sim\!C(=O)\!\sim \xleftarrow{\beta\text{-scission}} \sim\!C(\dot{O})\!\sim \xleftarrow{h\nu \text{ or } \Delta} \sim\!C(OOH)\!\sim$$

The solubility in toluene will obviously increase. The secondary hydrogen atoms, which are less labile than the tertiary atoms but far more numerous, are also abstracted and chain scission occurs thus explaining the decrease in mechanical properties beyond those of non-cross-linked polyethylene.

After photothermal or photocatalytic oxidation of a cross-linked polyethylene, the solubility in toluene does not increase, the Young's modulus increases and ultimate strength and elongation at break decrease (Fig. 11, curve (b)). These different relationships show that chain scissions occur without implying specifically the C—C bonds between two polymeric chains; the size of the macromolecule does not change, the insolubility is invariant. In photochemical initiation, abstraction of tertiary hydrogen atoms by the reactive species does not appear to be easier than abstraction of secondary hydrogen atoms.

2.4. Mechanism of Photocatalytic Oxidation of a Cross-linked Polyethylene

The results reported in the previous sections are summarised by the scheme presented in Fig. 12.

Hydrogen-bonded hydroperoxides are observed in the polymer at a very low concentration. In the liquid phase, at the same concentration, only monomeric forms of model compounds (i.e. t-butylhydroperoxide) appear in the IR spectrum. In polypropylene (PP),[38] hydroperoxides are formed as dimer, trimer or longer sequences.

The influence of hydroperoxides on the overall oxidation has been extensively studied. It is of great fundamental and practical importance to assess the existence of secondary photoinduction due to hydroperoxides. Such experiments have to be made in transparent samples in which the photostationary concentration of hydroperoxide is very low; higher concentrations of peroxides can be obtained by a thermal pre-oxidation of the sample. In the absence of oxygen, photolysis of tertiary and secondary hydroperoxides yield mainly alcohols and ketones respectively, in polyethylene maintained at 50 °C. In the presence of oxygen, photolysis of secondary hydroperoxide prepared in the polymer yields exactly the same quantities of ketones as in the absence of oxygen. Therefore, photolysis of hydroperoxide in the presence of oxygen does not induce any new formation of carbonyl compounds through an oxidation reaction initiated by radicals formed from the peroxides. Such induction does not occur at 50 °C. As hydroperoxides are certainly a source of free-radicals and as these radicals are reactive, the simplest way to understand the inefficiency of a new reaction is to suppose a fast reaction between the pairs of radicals formed:

$$\text{>C(H)–O–O–H} \xrightarrow[\Delta]{h\nu \text{ or}} \text{>C(H)–}\dot{\text{O}}\ \ \dot{\text{O}}\text{–H} \longrightarrow \text{>C=O} + H_2O$$

In this reaction it is assumed that the tertiary hydrogen of the macroalkoxy-radical is highly labile.

Petruj and Marchal[39] also assume a direct decomposition of secondary hydroperoxide into ketones through a chain mechanism in the thermal and radiochemical oxidation of polyethylene. The propagation steps are:

$$
\begin{array}{c} -CH_2 \\ | \\ H-C-OO\cdot \\ | \\ -CH_2 \end{array} + \begin{array}{c} -CH_2 \\ | \\ H-C-OOH \\ | \\ -CH_2 \end{array} \rightarrow \begin{array}{c} -CH_2 \\ | \\ H-C-OOH \\ | \\ -CH_2 \end{array} + \begin{array}{c} -CH_2 \\ | \\ \cdot C-OOH \\ | \\ -CH_2 \end{array}
$$

$$
\begin{array}{c} -CH_2 \\ | \\ \cdot C-OOH \\ | \\ -CH_2 \end{array} \rightarrow \begin{array}{c} -CH_2 \\ \diagdown \\ C{=}O \\ \diagup \\ -CH_2 \end{array} + OH\cdot
$$

$$
OH\cdot + \begin{array}{c} -CH_2 \\ | \\ H-C-OOH \\ | \\ -CH_2 \end{array} \rightarrow H_2O + \begin{array}{c} -CH_2 \\ | \\ H-C-OO\cdot \\ | \\ -CH_2 \end{array}
$$

In polypropylene, t-peroxide is formed and photodissociation yields two radicals which cannot destroy each other through hydrogen abstraction since no labile hydrogen is available.

$$
\begin{array}{c} \diagdown\ \diagup \\ C \\ \diagup\ \ \diagdown \\ O \qquad CH_3 \\ | \\ O \\ \diagdown \\ H \end{array} \xrightarrow[\Delta]{h\nu\ \text{or}} \begin{array}{c} \diagdown\ \diagup \\ C \\ \diagup\ \ \diagdown \\ \dot{O} \qquad CH_3 \end{array} + OH\cdot
$$

It is thus to be expected that hydroperoxide has an inductive role in the oxidation of polypropylene and, as shown later, this is indeed so.

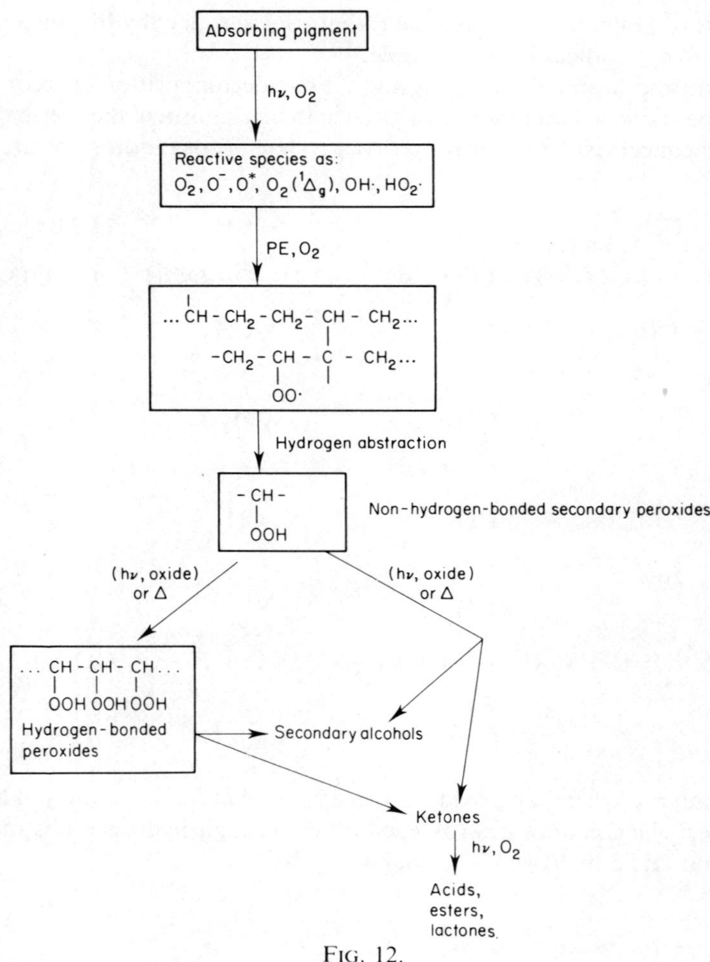

FIG. 12.

3. PHOTOCATALYTIC OXIDATION OF POLYETHYLENE

It is to be expected that hydrogen abstraction as the initiation process for photooxidation (Section 2.3.) should apply to 'high pressure, low density' polyethylene, both before and after cross-linking. We have indeed observed the same photoproducts to appear at the same rates in polyethylene as in cross-linked polyethylene, in the absence and in the presence of absorbing oxides. In polyethylene, only a small percentage of photocatalytic oxide ($<1\%$) can be added to the polymer; wavelengths in the range 300–380 nm are nevertheless totally absorbed by the pigment and photocatalysis is the major phenomenon.

The principal features of photocatalytic oxidation of polyethylene may be summarised as follows:

The primary photoproducts are secondary hydroperoxides which are thermally stable at room temperature and which can be photolysed in transparent material. Wavelengths up to 380 nm can lower the stationary concentration of hydroperoxide below that detectable by IR spectrophotometry.

In the presence of low contents of TiO_2 (anatase or rutile) or ZnO, photochemical protection of hydroperoxide is observed; hydroperoxides associate by hydrogen-bonding.

Secondary hydroperoxides which accumulate in either thermal or photocatalytic oxidation are not photoinducers. If irradiated, they decompose directly to ketones without inducing any new oxidation processes.

Ketones participate in secondary photochemical processes in Norrish Type I or Norrish Type II reactions and give rise to unsaturation and chain scissions.

No photostabilisation is observed on addition of triphenyl phosphite (0·2 and 0·5%). As hydroperoxides have no secondary inducting effect, such inefficiency is to be expected.

Thermal oxidation is inhibited by small amounts of phosphites (0·1%).

It has also been observed that addition of small amounts of carbon black (up to 1%) induces fast oxidation under u.v. irradiation; the maximum rate of photoinduced oxidation occurs with 0·3% carbon black.

4. PHOTOCATALYTIC OXIDATION OF POLYPROPYLENE

The photothermal oxidation of isotactic and atactic polypropylene has attracted much attention.[40–55] A detailed analysis of various aspects of this reaction has been made, especially by Carlsson and Wiles. The photocatalytic activity of pigments on polypropylene has been less extensively studied. Irich has found a correlation between photoactivity of metal oxides such as TiO_2 (anatase and rutile), ZnO and even $BaSO_4$† in

† We have shown in preliminary studies that $BaSO_4$ and $CaCO_3$ have no influence on the photooxidation of a cross-linked polyethylene.

polypropylene photooxidation and quantum yields in the oxidation of isopropanol sensitised by these pigments.[13] Recently, McKellar *et al.* have also reported quenching by TiO_2 anatase and not by rutile, of the phosphorescence of α, β-unsaturated ketones, present as impurities in commercial polypropylene. These impurities have been proposed as the initial inducers of photooxidation.[15]

A study of the photocatalytic oxidation of polypropylene is in progress in our research group.[56] Two main points will be discussed:

First, a comparison will be made of photoinduction by hydroperoxides in polypropylene and in polyethylene. As pointed out in previous sections, secondary hydroperoxides formed in the primary steps of oxidation of polyethylene have no inductive effect. On the other hand, it is to be expected that tertiary hydroperoxides formed in the oxidation of polypropylene will have a marked inductive effect under the same experimental conditions as for polyethylene. Carlsson and Wiles have already reported initiation by photodecomposition of hydroperoxide.[40,51]
Second, the influence of zinc oxide (365·5–Societe Vieille Montagne) on the course of photooxidation of isotactic polypropylene.

4.1. Photoinduction by Hydroperoxides

In transparent isotactic or atactic polypropylene (ICI Ltd) the low photostationary concentrations of hydroperoxides are beyond detection by IR spectrophotometry. Changes in absorbance at $1712\,cm^{-1}$ and $3400\,cm^{-1}$ of isotactic and atactic polypropylene under irradiation are represented in Fig. 13. Self-acceleration is observed only with the atactic polymer. The IR absorption at $3400\,cm^{-1}$ must be attributed to associated alcohols, since on photolysis of photooxidised samples in vacuum no decrease in absorption is observed. Treatment with an alcoholic solution of sodium hydroxide shows that most of the absorption at $1712\,cm^{-1}$ is due to acids, even in the earlier stages of the oxidation.

Accumulation of tertiary hydroperoxides can be observed in the IR spectra on thermal oxidation at 130 °C. A relatively sharp absorption band at $3550\,cm^{-1}$ is attributed to free hydroperoxides and an unstructured band around $3400\,cm^{-1}$ is associated with hydrogen-bonded hydroperoxide (Fig. 14). If this pre-oxidised sample is irradiated with polychromatic light ($\lambda > 305$ nm) for a short time in vacuum or in the presence of oxygen, the IR spectrum is modified as shown in Fig. 14. In vacuum, hydroperoxides and ketones are photolysed, and unsaturation appears as a result of Norrish Type II processes. In the presence of oxygen, hydroperoxides are photolysed and induce a fast oxidation of polypropylene to produce

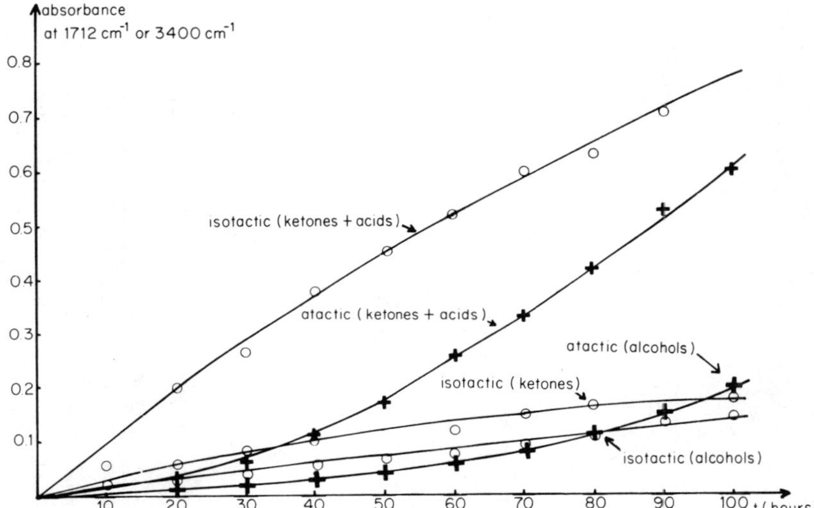

FIG. 13. Photothermal oxidation of atactic and isotactic polypropylene; ○, Changes in absorbance at $1712\,cm^{-1}$; +, Changes in absorbance at $3400\,cm^{-1}$; polychromatic light $\lambda > 305\,nm$; temperature of sample, 50 °C.

ketones. Meanwhile, ketones are also photolysed and unsaturation appears. In a steady photothermal oxidation, during the same short period and at the same initial content of carbonyl compounds, absorbance at $1712\,cm^{-1}$ remains practically invariant (i.e. $\Delta D < 0{\cdot}05$ in 5 h). It can be concluded that by thermal- or photodissociation of tertiary hydroperoxides, radicals are formed which are able to initiate new oxidation processes in experimental conditions in which secondary hydroperoxides formed in polyethylene are ineffectual.

$$-CH_2-\underset{\displaystyle OOH}{\overset{\displaystyle CH_3}{\overset{|}{\underset{|}{C}}}}-CH_2- \xrightarrow[\Delta]{h\nu\ \text{or}} -CH_2-\underset{\displaystyle \dot{O}}{\overset{\displaystyle CH_3}{\overset{|}{\underset{|}{C}}}}-CH_2- + OH\cdot$$

β-scission; PP, O_2; PP, O_2

$$-CH_2-\underset{\displaystyle \backslash\backslash O}{\overset{\displaystyle CH_3 /}{C}} + \cdot CH_2\sim \xrightarrow{PP,\ O_2} \text{oxidation chain reactions}$$

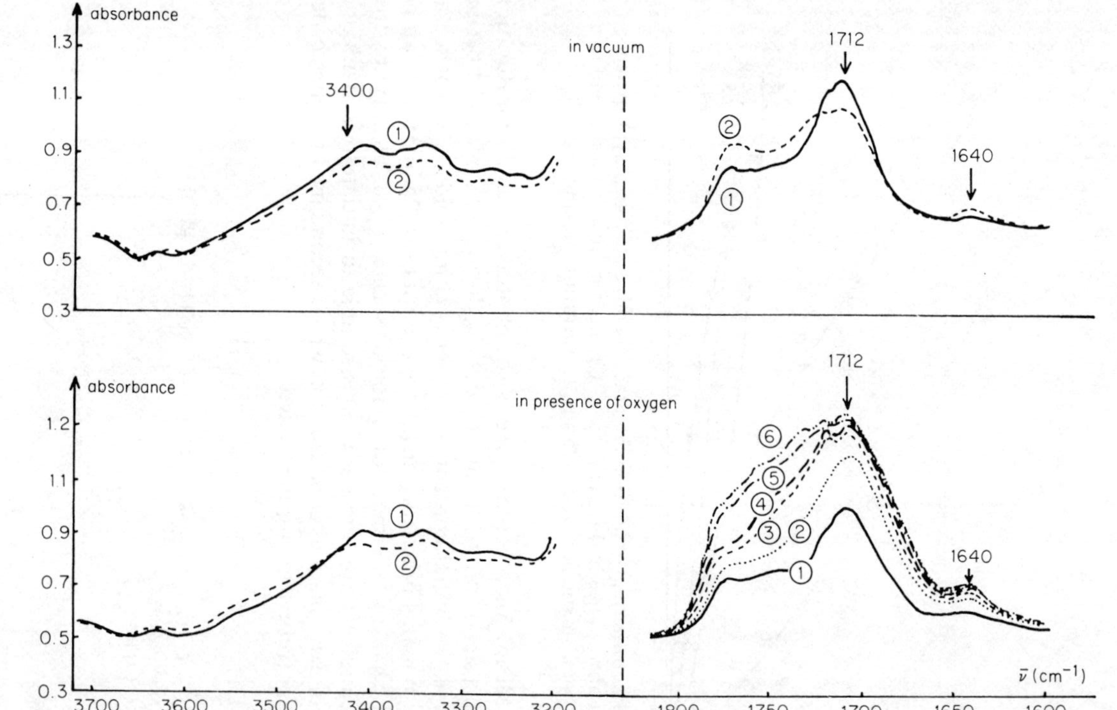

FIG. 14. Changes in the IR spectrum of a thermally oxidised sample of isotactic propylene under irradiation in vacuum and in the presence of oxygen; pre-thermal oxidation at 130°C; polychromatic light $\lambda > 305$ nm; temperature of sample, 60°C.

The photolysis of tertiary hydroperoxides gives rise to tertiary alcohols, and, through β-scission of macroalkoxy-radicals, to ketones.

4.2. Zinc Oxide Photocatalysed Oxidation of Isotactic Polypropylene

The addition of 3–10% of zinc oxide (365·5–Societe Vieille Montagne) to isotactic polypropylene has three consequences:

First, the evolution of IR spectra is quite different in photocatalytic oxidation compared with photothermal oxidation. Figure 15 shows the variations of absorbance at 1712 cm^{-1} and at 3400 cm^{-1} in both cases. For the same carbonyl content, the hydroxyl absorption is far higher in the

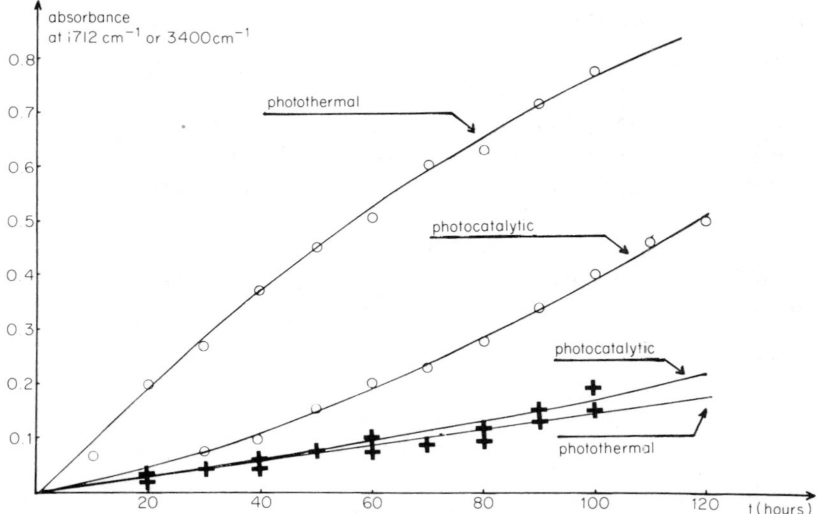

FIG. 15. Comparison of the photocatalytic and photothermal oxidation of isotactic polypropylene; ○, Changes in absorbance at 1712 cm^{-1}; +, Changes in absorbance at 3400 cm^{-1}; polychromatic light $\lambda > 305$ nm; temperature of sample, 50 °C.

presence of ZnO. The unstructured absorption around 3400 cm^{-1} must be attributed to hydrogen-bonded hydroperoxides; no absorption around 3550 cm^{-1} corresponding to free hydroperoxides can be observed. In the range 1800–1650 cm^{-1}, no significant differences appear in the complex carbonyl absorption. The addition of ZnO inhibits the formation of unsaturation at 1640 cm^{-1}.

Second, the rate of formation of carbonyl compounds absorbing at $1712\,cm^{-1}$ (ketones + acids) is, in the first stage of the reaction, inhibited by ZnO. As wavelengths in the range 300–400 nm are totally absorbed by the pigments, this means that in photocatalytic oxidation, the quantum yield for the appearance of carbonyl compounds is far lower than the same quantum yield in photothermal oxidation. Comparison of the photo-oxidation of polyethylene, cross-linked or not, and of polypropylene under the same experimental conditions (50 °C, polychromatic light $\lambda > 305$ nm) shows that the initial rate of appearance of carbonyl compounds is, at most, three times as high in polypropylene. Zinc oxide may act as a protecting agent in polypropylene and as a degrading additive in polyethylene.

Third, self-acceleration of the formation of carbonyl compounds occurs only in photocatalytic oxidation.

These analytical and kinetic observations are easily explained in terms of the accumulation of hydrogen-bonded hydroperoxide in the presence of pigments. The absorbing oxide affords photochemical protection for hydroperoxide as well as for ketones. Any secondary step, such as photolysis of hydroperoxide or Norrish Type II processes in ketones, is then inhibited.

In photocatalytic oxidation the concentration of hydroperoxide is controlled by photochemical formation and thermal disappearance. When hydroperoxide attains a critical concentration, thermal induction appears and self-acceleration might thus be explained.

ACKNOWLEDGEMENT

This work was supported by the Délégation Générale à la Recherche Scientifique et Technique—Secrétariat d'Etat à la Recherche.

REFERENCES

1. Steinbach, F. and Barth, R., *Ber. Phys. Chem.*, **73** (1969), p. 884.
2. Steinbach, F. and Muller, H. D., *Z. Phys. Chem.*, **84** (1973), p. 277.
3. Steinbach, F. and Haborth, R., *Faraday Disc. Chem. Soc.*, **58** (1974) p. 143.
4. Formenti, M., Juillet, F., Meriadeau, P. and Teichner, S. J., *Chem. Technol.*, **1** (1971) p. 680; *Bull. Soc. Chim. Fr.* (1972) p. 69.
5. Djeghri, N., Formenti, M., Juillet, F. and Teichner, S. J., *Faraday Disc. Chem. Soc.*, **58** (1974) p. 185.

6. Bourasseau, S., Martin, J. R., Juillet, F. and Teichner, S. J., *J. Chim. Phys.*, **71** (1974) p. 1017; Ibid **71** (1974) p. 1025.
7. Formenti, M., Juillet, F. and Teichner, S. J., *Bull. Soc. Chim. Fr.* (1976) pp. 1031, 1315.
8. Courbon, H. and Pichat, P., *Comptes Rendus Acad. Sc. Paris*, **285** (1977) p. 171.
Courbon, H., Formenti, M. and Pichat, P., *J. Phys. Chem.*, **81** (1977) p. 550.
9. Herrmann, J. M., Disdier, J. and Pichat, P., Proc. 3rd Intern. Conf. Solid Surface, Vienna, 1977.
10. Mozzanega, M. N., Herrmann, J. M. and Pichat, P., *Tetrahedron Letters* (1977) p. 2965.
11. Hata, K., Kawasaki, S., Kubokawa, Y. and Miyata, H., Proc. 6th Int. Cong. Catal. 1976 (C. G. Bond, P. B. Wells, F. C. Tompkins, Eds.), England, Chemical Society.
12. Bickley, R. I. and Jayanti, R. K. M., *Faraday Disc. Chem. Soc.*, **58** (1974) p. 194.
13. Irich, G., *J. Appl. Polym. Sci.*, **16** (1972) p. 2387.
14. Pappas, S. P. and Fischer, R. M., *J. Paint Technol.*, **46** (1974) p. 65.
15. Allen, N. S., McKellar, J. F., Philipps, G. O. and Wood, D. G. M., *J. Polym. Sci., Chem. Ed.*, **12** (1974) p. 241.
16. Allen, N. S., McKellar, J. F., Philipps, G. O. and Chapman, C. B., *J. Polym. Sci., Chem. Ed.*, **12** (1974) p. 723.
17. Laurenson, P., Arnaud, R., Lemaire, J., Quemner, J. and Roche, G., *Europ. Polym. J.*, **14** (1978) p. 129.
18. Boasson, E. H., Kamerbeck, B., Algera, A. and Kroes, G. H., *Rec. Trav. Chim. Pays-Bas*, **81** (1962) p. 625.
19. Kroes, G. H., *Rec. Trav. Chim. Pays-Bas*, **82** (1963) p. 979.
20. Egerton, G. S., *Nature*, **204** (1964) p. 1153.
21. Egerton, G. S. and Shah, K. M., *Text. Res. J.*, **38** (1968) p. 130.
22. Taylor, H. A., Tincher, W. C. and Hammer, W. F., *J. Appl. Polym. Sci.*, **14** (1970) p. 141.
23. Allen, N. S., McKellar, J. F. and Wood, D. G. M., *J. Polym. Sci., Chem. Ed.*, **13** (1975) p. 2319.
24. Ranby, B. and Rabek, J. F., *Photodegradation, photooxidation and photostabilization of polymers*, New York, John Wiley, 1975, p. 122ff.
25. Adams, J. H., *J. Polym. Sci., Part A*1, **8** (1970) p. 1279.
26. Ranby, B. and Rabek, J. F., *Photodegradation, photooxidation and photostabilization of polymers*, New York, John Wiley, 1975, p. 306.
27. Ronayette, J., Arnaud, R., Lebourgeois, P. and Lemaire, J., *Can. J. Chem.*, **52** (1974) pp. 1848, 1858.
28. Gardette, J.-L., Guyot, G., Arnaud, R. and Lemaire, J., *Nouveau Journal de Chimie*, **1** (1977) p. 287.
29. Hatchard, G. and Parker, C. A., *Proc. Roy. Soc. London*, **A220** (1953) p. 104, *Proc. Roy. Soc. London*, **A235** (1956) p. 518.
30. Calvert, J. G. and Pitts, J. N., Jr, *Photochemistry*, New York, John Wiley, 1968, p. 783.
31. Bowman, W. D. and Demas, J. N., *J. Phys. Chem.*, **80** (1976) p. 2434.
32. Pappas, S. P. and Kuhhirt, W., *J. Paint Tech.*, **47** (1975) p. 42.

33. WICKS, Z. W., Jr and KUHHIRT, W., *J. Paint Tech.*, **47** (1975) p. 50.
34. NIKI, E., DECKER, C. and MAYO, F. R., *J. Polym. Sci., Chem. Ed.*, **11** (1973) p. 2813.
35. KATO, Y., CARLSSON, D. J. and WILES, D. M., *J. Appl. Polym. Sci.*, **13** (1969) p. 1447.
36. DECKER, C. D., MAYO, F. R. and RICHARDSON, H., *J. Polym. Sci., Chem. Ed.*, **11** (1973) p. 2879.
37. ARNAUD, R., LEMAIRE, J., QUEMNER, J. and ROCHE, G., *Europ. Polym. J.*, **12** (1976) p. 499.
38. CHIEN, J. C. W., VANDENBERG, E. J. and JABLONER, H., *J. Polym. Sci.*, **6** (1968) p. 381.
39. PETRUJ, J. and MARCHAL, J., XIIth Annual French–Techcoslovak Meeting on Ageing of Polymers, Novy Smokobec, 1978 (unpublished).
40. CARLSSON, D. J. and WILES, D. M., *Macromol.*, **2** (1969) p. 597.
41. HARPER, D. J. and MCKELLAR, J. F., *Chem. Ind.* (1972) p. 848.
42. JIRACKOVA, L. and POSPISIL, J., *Europ. Polym. J.*, **9** (1973) p. 71.
43. MILL, T., RICHARDSON, H. and MAYO, F. R., *J. Polym. Sci., Chem. Ed.*, **11** (1973) p. 2899.
44. DECKER, C. and MAYO, F. R., *J. Polym. Sci., Chem. Ed.*, **11** (1973) p. 2847.
45. CARLSSON, D. J. and WILES, D. M., *J. Polym. Sci., Chem. Ed.*, **11** (1973) p. 759.
46. CARLSSON, D. J. and WILES, D. M., *Macromol.*, **7** (1974) p. 259.
47. RANBY, B. and RABEK, J. F., Photodegradation, photooxidation and photostabilization of polymers, New York, John Wiley, 1975, p. 128.
48. ASPLER, J., CARLSSON, D. J. and WILES, D. M., *Macromol.*, **9** (1976) p. 691.
49. CARLSSON, D. J., GARTON, A. and WILES, D. M., *Macromol.*, **9** (1976) p. 695.
50. CARLSSON, D. J. and WILES, D. M., *J. Macromol. Sci., Rev. Macromol. Chem.*, **C14** (1976) p. 65.
51. CARLSSON, D. J. and WILES, D. M., Ultraviolet Light Induced Reactions in Polymers (S. S. Labana, Ed.), *Am. Chem. Soc. Symp. Ser.*, **25** (1976), p. 321.
52. CARLSSON, D. J. and WILES, D. M., *J. Macromol. Sci., Rev. Macromol. Chem.*, **C14** (1976) p. 155.
53. ALLEN, N. S., MCKELLAR, J. F. and PHILIPPS, G. O., *J. Polym. Sci., Chem. Ed.*, **12** (1974) p. 2647.
54. ALLEN, N. S., HOMER, J. and MCKELLAR, J. F., *J. Appl. Polym. Sci.*, **20** (1976) p. 2553.
55. ALLEN, N. S., MCKELLAR, J. F. and PHILIPPS, G. O., *Am. Chem. Soc. Polymer Preprints*, **18** (1977) p. 375.
56. GINHAC, J. M., *Thèse de Doctorat de* 3° *Cycle*, University Clermont II, France, in preparation.

Chapter 7

THERMO-OXIDATIVE DEGRADATION OF POLY(VINYL-CHLORIDE)

F. TÜDŐS, T. KELEN and T. T. NAGY

Central Research Institute for Chemistry of the Hungarian Academy of Sciences, Budapest, Hungary

SUMMARY

The main features of the thermo-oxidative degradation of PVC compared to the pure thermal process and the most commonly used methods for its investigation are briefly reviewed. The nature of pure thermal degradation, especially primary HCl loss and the formation of polyene sequences, are outlined. The kinetics and mechanism of oxidation of polyene sequences is discussed in detail. The oxidisability of polyenes is extremely high and approximately proportional to sequence length. Mainly cyclic peroxides, but also hydroperoxides, are formed in the autooxidation of polyenes. Hydroperoxides are the predominant initiating species.

Polyene content greatly enhances HCl loss in the early stages of thermo-oxidative degradation. The most significant processes in the low conversion range are outlined. Thermally initiated HCl loss, oxidation of polyene sequences and initiation of further HCl loss by radicals attacking intact monomer units form a complicated reaction system, involving consecutive, competitive and 'feedback' steps.

The kinetics of HCl loss is presented for the thermo-oxidative degradation to high conversions of PVC powder and solution. Auto-acceleration is present in powder but absent in solution. Quantitative data are given about chain scission and cross-linking. Conclusions regarding thermo-oxidative degradation and stabilisation are drawn and some questions, which are still unresolved, are raised.

1. INTRODUCTION

Over the past few years extensive studies have been made on PVC degradation.[1-7] Thermo-oxidative processes have not received serious consideration, attention having been focused mainly on non-oxidative 'pure' thermal degradation and stabilisation, in spite of the fact that the rate of evolution of HCl is considerably higher in the presence of oxygen than in an inert atmosphere or in vacuum (see, for example, references 8–10).

Although, quite obviously the effect of oxygen on degradation cannot be considered as a simple catalytic process, it is occasionally referred to in this way in the literature.[11-12] Thermo-oxidative degradation of solid PVC samples is an accelerating process. This is especially pronounced at high temperatures (above 180 °C) and high extents of HCl loss.[11,13,14] Under these conditions, however, auto-acceleration can also be observed in inert atmospheres.[15-18] In spite of this, some authors consider auto-acceleration to be an inherent feature of thermo-oxidative degradation.[11,19]

A characteristic difference may be discerned in the colour and u.v.-visible spectra of samples degraded in inert atmospheres and those degraded in oxygen. At identical extents of HCl loss, PVC degraded in the presence of oxygen has a lighter colour. Absorption maxima corresponding to polyenes of different lengths, easily visible in the spectrum of thermally degraded PVC, cannot be observed in the case of oxidative degradation which gives unresolved patterns.[20,21]

In the course of thermal degradation, practically no scission of the main chain occurs,[22,23] whereas in the presence of oxygen, chain scissions play an important role.[9,11]

Thermo-oxidative degradation has usually been followed by the same methods as for thermal degradation. Elimination of HCl is usually followed by continuous measurements and several authors have also investigated the changes of u.v. and visible spectra. Procedures allowing simultaneous observation of the various parameters in the course of degradation have proved to be very useful. Guyot and Benevise[24] measured the weight of the sample and evolved HCl. Wolkóber *et al.* constructed an instrument (Dinamoxmeter) applicable for simultaneous measurement of oxygen uptake and HCl loss in thermo-oxidative and photooxidative degradation.[25-26] Chain scission and cross-linking in the course of thermo-oxidative degradation was followed mainly by intrinsic viscosity and gel fraction measurements,[9,27,28] but absolute molecular weight measurements[29] as well as the gel permeation chromatography (GPC) method[22] were also applied. It has been shown by IR spectroscopy and chemical transformation of the polymer that most of the carbonyl groups are conjugated with double bonds.[30,31]

There are various interpretations of the mechanism of thermo-oxidative degradation.[19,12–14,19,21,32–37] It is generally accepted that in the presence of oxygen, radical chain reactions play an important role. This is supported by the observation that the high rate of HCl elimination in thermo-oxidation is reduced by antioxidants.[38,39] The u.v.-visible spectra of PVC degraded in air in the presence of an antioxidant are similar to the spectra of samples degraded in inert atmospheres: both contain easily discernable peaks and shoulders which may be assigned to polyenes of various lengths.[40] Several authors have also found the rate of PVC degradation to be increased in inert atmospheres by radical initiators,[41–44] contrary to the findings of Braun and Bender.[10] The degradation of PVC containing peroxide groups formed by pre-treatment with ozone[42,45,46] or due to the presence of oxygen during polymerisation,[47] was found to be faster than the degradation of the untreated sample.

In this chapter we shall summarise our ideas on the thermo-oxidative degradation of unstabilised PVC. In some essential points we agree with the findings of Rieche *et al.*[33] and Braun,[21] but we also report some new aspects and a wider range of experimental evidence. The reader is referred to recent reviews and/or publications, e.g., in the field of thermal degradation of PVC,[1–7,48,49] stabilisation of PVC in the presence of oxygen,[38–40,50] photooxidative degradation of PVC,[51] pyrolysis of PVC (above 250 °C) in oxidative atmospheres[52,53] and autooxidation of hydrocarbons and unsaturated hydrocarbons.[54,55]

For a better understanding of the thermo-oxidative degradation of PVC, knowledge of the kinetics and mechanism of polyene oxidation and pure thermal degradation is required. These are rather complex processes which have the advantage, however, that they can be studied separately from the even more complicated process of thermo-oxidative degradation.

2. THERMAL DEGRADATION OF PVC

In the pure thermal degradation of PVC, HCl elimination occurs in a series of allyl-activated steps (zip reaction). Conjugated polyenes of various length are formed:

```
        CH     CH2    CH2
   \  //  \   /  \   /  \   /
    CH     CH     CH     CH
           |      |      |
           Cl     Cl     Cl

              | −HCl
              ↓

        CH     CH     CH2
   \  //  \  //  \   /  \   /
    CH     CH     CH     CH
                  |      |
                  Cl     Cl

              | −HCl
              ↓

        CH     CH     CH
   \  //  \  //  \  //  \   /
    CH     CH     CH     CH
                         |
                         Cl

              ↓

             etc.
```

The polyene sequences are short, the average length usually being in the range 4–10, which means that instead of proceeding along the entire molecule, the allyl-activated chain reaction is terminated.[20,21,48] In the primary HCl elimination process the following three basic steps may be distinguished:[48]

(1) Initiation of HCl unzipping.
(2) Allyl-activated chain propagation.
(3) Chain termination.

Initiation may occur at weak sites in the polymer or at random at a lower rate.[6,48] The rate constant of chain termination is high so that the rate of degradation soon reaches a steady-state and the concentration of intermediates (still growing polyenes) is rather low. The mechanism of chain termination is not completely clear, e.g., ring-closure may occur[6,48,49] at the growing end of the polyene, or the lengths of the polyenes are partly or fully determined by *a priori* given sequences.[56] If the probability of chain termination is independent of the length of the growing polyene, there will be a continuous distribution of polyene lengths. In our experience the concentration of shorter polyenes is higher than that of longer polyenes[21] and the distribution is continuous.[48]

Primary HCl loss and polyene formation are accompanied by secondary processes,[57] e.g. cyclisation of polyene sequences,[49,58] formation of a small amount of benzene[6,21,49] and cross-linking when the PVC is degraded in the solid phase.[9,22,28,59]

3. THERMAL OXIDATION OF POLYENES

Unsaturated hydrocarbons, especially conjugated dienes and polyenes, are known to be very sensitive to oxidation.[54,55] The two main and often competing propagation steps in their radical oxidation are the addition of a peroxyradical to the double bond and the abstraction of an allylic H atom.[60] There is no comprehensive knowledge of the oxidation of long, conjugated polyenes; the only exception seems to be β-carotene which has 11 conjugated double bonds.[61,62] The bleaching effect of oxygen on degraded PVC has long been known and was shown to be caused by the radical oxidation of long polyenes.[63,64] Thermally degraded PVC contains polyenes of different sequence lengths and also 1,3-cyclohexadiene groups,[49,58] which can be formed in several possible intermolecular propagation steps e.g.:

ROO˙ + (polyene) → ROO-adduct radical (two isomers) ; → allylic radical + ROOH

ROO˙ + (cyclohexadiene) → (cyclohexadienyl radical) + ROOH

In the presence of oxygen, carbon radicals are immediately converted to peroxyradicals. Isomeric peroxyradicals can be formed from a given delocalised polyenyl radical, which greatly increases the number of possible products:

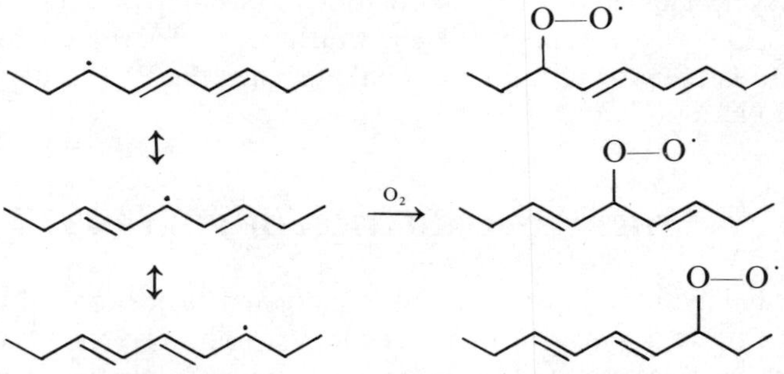

In the oxidation of polyenes a distinction should be made between intermolecular and intramolecular chain propagation steps. In the former, oxidative chain-propagation proceeds from one polyene to another, i.e. the polyene is attacked by the peroxyradical as shown above, whereas in the latter case the peroxyradical reacts with a nearby double bond in the same molecule thus forming cyclic peroxides, most probably 6- and 5-membered rings:[65]

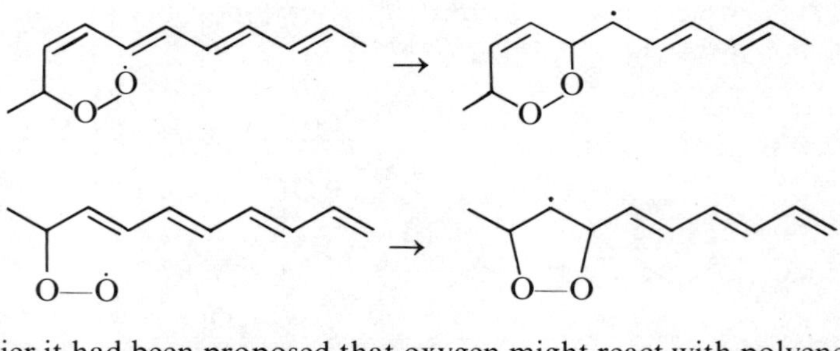

Earlier it had been proposed that oxygen might react with polyenes as a diene reagent,[32,33,39] to form similar cyclic peroxides. This reaction is, however, spin-forbidden with ground-state triplet oxygen molecules. Since such products can be produced only with singlet oxygen,[66] this reaction plays little or no part in thermo-oxidative PVC degradation:

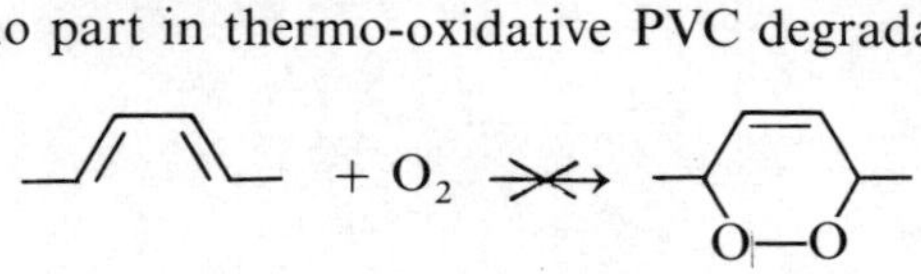

The oxidation of the polyene sequences in thermally degraded PVC is a rather complicated chain process. To gain a better insight into the mechanism and to avoid complications in the kinetics one must ensure well defined reaction conditions. The consumption of the individual polyenes can be followed by u.v.-visible spectrometry.[58,59,65,67,68] (Figure 1; short polyenes, containing less than 5 conjugated double bonds could not be measured, due to the u.v. absorbance of the solvent.)

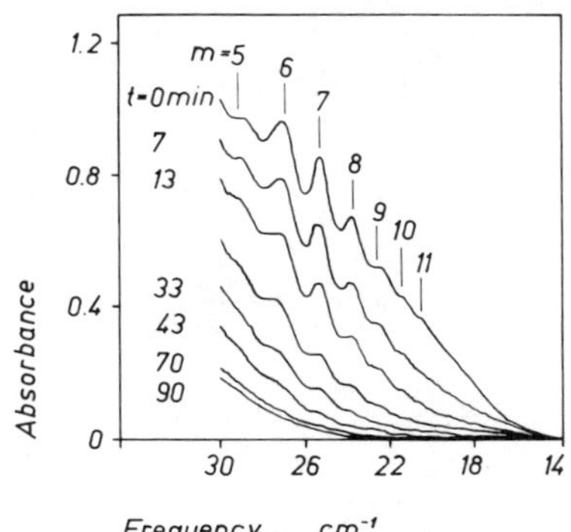

FIG. 1. Changes in the u.v. and visible spectra of thermally degraded PVC during initiated oxidation. Experimental conditions: PVC predegraded in 10 g/litre solution in 1,2,4-trichlorobenzene (TCB) at 210 °C in Ar, to 1 % HCl loss; oxidised at 60 °C in 1 atm O_2 with 1·3 mmol/litre azo-bis-isobutyronitrile. Photometry in 1:1 mixture of TCB and tetrahydrofuran solution (5 g/litre). *m* Is the polyene length, i.e. the number of conjugated double bonds in the polyene.

Surprisingly simple oxidation kinetics were observed when experiments were carried out in pure oxygen at atmospheric pressure in 1,2,4-trichlorobenzene as solvent, using a suitable free-radical initiator in order to keep the rate of initiation constant.[65] In such cases oxidation obeys a first order rate law, i.e., the log $[(A - A_\infty)/(A_0 - A_\infty)]$ versus time plot for a given polyene length is linear. The slope of such plots, i.e. the rate constant for the consumption of polyene sequences turned out to be proportional to the polyene length.

This is illustrated in Fig. 2 where the product of polyene length and oxidation time ($m \times t$) is used as the independent variable. All points are seen to lie on the same straight line. Thus, at a given initiator concentration the reaction rate-constant can be characterised by a single value, namely slope *W* of this straight line, which may be considered as the overall oxidation rate. When the experimental conditions are changed, behaviour typical of radical oxidation processes is observed. Thus *W* is proportional to the square root of the initiator concentration, indicating bimolecular termination, and it is practically independent of partial oxygen pressure above 0·5 atm. This is explained by fast conversion of alkyl (i.e. polyenyl) radicals into peroxyradicals, the latter predominantly participating in chain termination.

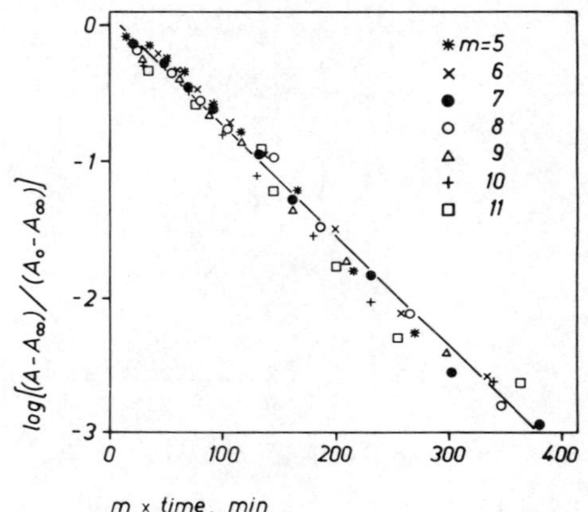

FIG. 2. Logarithmic plot of polyene consumption as a function of the product of polyene length and oxidation time. Experimental conditions as in Fig. 1 (A = absorbance).

In the description of the kinetics of polyene oxidation, the formation of shorter from longer polyenes should be taken into account. Their effect in the evaluation of results is, however, negligible due to intramolecular propagation reactions. One would expect more complicated kinetics if this secondary formation of shorter polyenes were not negligible. When the initiator (azo-bis-isobutyronitrile) reacts with the polyenes in an inert atmosphere, polyene consumption is much slower and the kinetics are complicated.[65] At a sufficient partial pressure of oxygen, the polyene sequence is practically consumed after attack by a peroxyradical in intermolecular propagation. The proportionality of polyene consumption-rate to polyene length (Fig. 2) indicates that the rate constant of intermolecular propagation must be proportional to the polyene length, at least for polyene sequences longer than 4. In a chemical sense this means that the peroxyradical will react with any double bond of the polyene sequence with the same probability. In addition to these types of intermolecular propagation, hydrogen abstraction must also take place as a side reaction.

With the usual approximations applied for radical chain reactions, a kinetic equation for the initiated oxidation of polyene sequences has been deduced:[65]

$$\log(p_{m0}/p_m) = W(m \times t) \quad (1)$$

$$W = (2k_1 f)^{1/2} I^{1/2}[k_2/(2k_4)^{1/2}]$$

where W is the 'overall rate' of polyene consumption; p_{m0} and p_m are concentrations of the polyene sequence of length m at the beginning of oxidation and after time t respectively; k_l is the rate constant of the decomposition of the initiator; f is the initiation efficiency; k_2 is the rate constant for intermolecular propagation related to one double bond in the polyene sequence; k_4 is the rate constant for termination, and I the initiator concentration.

Using eqn. (1) the kinetic parameters of oxidation were calculated in the temperature range 40–70 °C. Oxidisability of polyenes ($k_2/(2k_4)^{1/2}$) could be expressed by the following Arrhenius equation:[65]

$$k_2/(2k_4)^{1/2} = 9{\cdot}8 \times 10^4 \exp[(-6300\,\text{cal/mol})/RT] \quad (\text{mol/litre})^{-1/2}\,\text{min}^{-1/2} \quad (2)$$

(Where $R = 1{\cdot}987$ cal/mol/K and $T =$ absolute temperature (K).) The autooxidation of polyene sequences in solution is similar to that of initiated oxidation, the rate of initiation however changes with oxidation time due to changes in the concentration of the initiating species, in other words due to degenerate chain branching. Fortunately proportionality between consumption rate and polyene length exists to a good approximation also in this case. At the beginning of the oxidation the rate increases with increasing oxidation time, but at later stages it decreases approximately according to the first order rate law (Fig. 3).[67]

The overall rate of polyene consumption is proportional to the partial pressure of oxygen at low oxygen pressures, but does not change much above 0·7 atm. The first order rate of the second stage (the W slope of the straight line in the $\log[(A - A_\infty)/(A_0 - A_\infty)]$ versus $m \times t$ plots) was found to be proportional to the square root of the initial double bond concentration. The overall activation energy was around 10 kcal/mol. The kinetics of the autooxidation of β-carotene is in every respect analogous to these findings.[61,62] The amount of hydroperoxides as well as the sum of hydroperoxide and dialkyl type peroxide concentrations can be measured selectively.[67,69] In the course of autooxidation of polyene sequences the total amount of peroxide groups increases gradually, whereas the

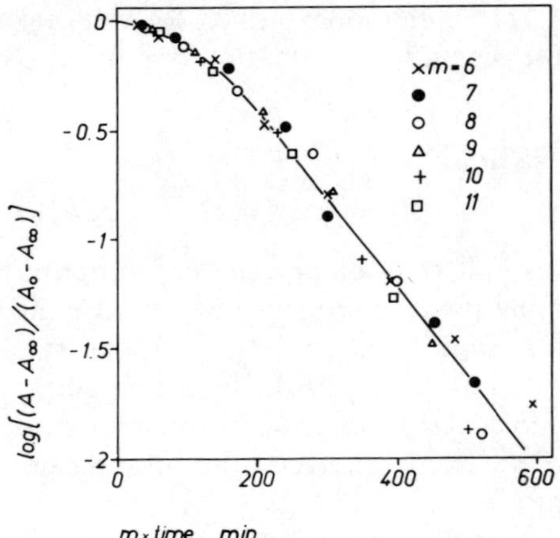

FIG. 3. Logarithmic plot of polyene consumption in autooxidation as a function of the product of polyene length and oxidation time. Experimental conditions: PVC predegraded as in Fig. 1; oxidised at 100 °C in 1 atm O_2.

concentration of hydroperoxides reaches a maximum shortly after the onset of oxidation, then decreases very slowly and is almost constant for a long period. The amount of dialkyl type, relatively stable cyclic peroxides is considerably higher than the concentration of hydroperoxides. At the end of oxidation the total amount of peroxide groups exceeds one quarter of the double bonds initially present in the reaction (Fig. 4).

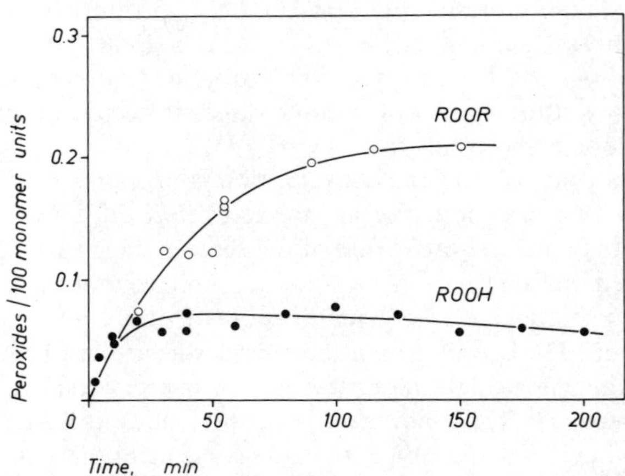

FIG. 4. Formation of peroxide during autooxidation of thermally degraded PVC. Experimental conditions as in Fig. 3.

The kinetics of polyene autooxidation can be explained by the kinetics of peroxide formation with the assumption that initiation is mainly caused by hydroperoxides. The initial increase in the rate of polyene consumption corresponds to increasing hydroperoxide concentration whereas in the second stage the first order polyene consumption corresponds to a constant rate of initiation, i.e. a nearly constant hydroperoxide concentration. Thus for this stage, similar to eqn. (1), we have:

$$W = (2k_3 f)^{1/2} \times [\mathrm{ROOH}]^{1/2} \times k_2/(2k_4)^{1/2} \qquad (3)$$

where k_3 is the rate constant of hydroperoxide decomposition, and [ROOH] denotes hydroperoxide concentration. The other symbols are as in eqn. (1). The proportionality found between W and the square root of initial double bond concentration is in line with eqn. (3) if hydroperoxide concentration in this stage is proportional to initial double bond concentration. The product of the rate constant and initiation efficiency of hydroperoxide decomposition was determined from the experimental data by means of eqn. (3) and was found to be $2k_3 f = 6 \,.\, 10^{-4}\ \mathrm{min}^{-1}$ at 100 °C.

The main features of the oxidation of polyenes in solution may be summarised as follows:

(1) If no initiator is added, initiation is caused mainly by hydroperoxide decomposition.
(2) Besides hydroperoxides, large amounts of cyclic peroxides are formed in intramolecular steps.
(3) The rate of intermolecular propagation is proportional to polyene length.
(4) At sufficiently high oxygen concentrations termination by two peroxyradicals predominates.

Solid phase autooxidation of polyene sequences in thermally degraded PVC has many features in common with autooxidation in the liquid phase.[68] Only the main differences are mentioned here. Although oxidation of longer polyenes is faster than of shorter ones in this case considerable deviations from proportionality between rate and sequence length can be observed. After a short induction period, the log $[(A - A_\infty)/(A_0 - A_\infty)]$ versus $m \times t$ plots show linear dependence, indicating a first order polyene consumption rate. Later on, the rate decreases markedly. The overall activation energy (about 27 kcal/mol) is higher than in the liquid phase. The dependence of the rate of polyene consumption on the partial pressure of oxygen is also different from that in the liquid phase. The rate in the solid phase is almost proportional to the partial pressure of oxygen, even in the region of 1 atm. These differences are most probably due to the lower mobility of reactants in the solid phase, but it is difficult to give even a qualitive explanation of these findings.

4. INITIAL STAGE OF THERMO-OXIDATIVE DEGRADATION OF PVC

From a practical point of view, the initial stages of degradation are the most important because PVC becomes a useless discoloured material even after only a few tenths of a percent of HCl loss. The initial phase of degradation is also decisive for a better understanding of the whole process.

Loss of HCl is measurable even in inert atmospheres at temperatures as low as 100 °C.[70] In oxygen, also, measurable but very slow HCl loss was detected at similar temperatures from virgin PVC powder.[68] In comparison to virgin PVC, HCl elimination was found to take place at a much higher rate in thermally pre-degraded (polyene-containing) samples (Fig. 5). Thus polyene oxidation is accompanied by considerable HCl loss. A similar observation was reported by Talamini and Pezzin, at 200 °C.[13] The rate of HCl evolution of pre-degraded samples passes through a maximum and does not reach a steady-state at 110 °C even in 300 min. In line with the last statement the estimated average lifetime of hydroperoxides in liquid-phase oxidation at this temperature is several hundred minutes.

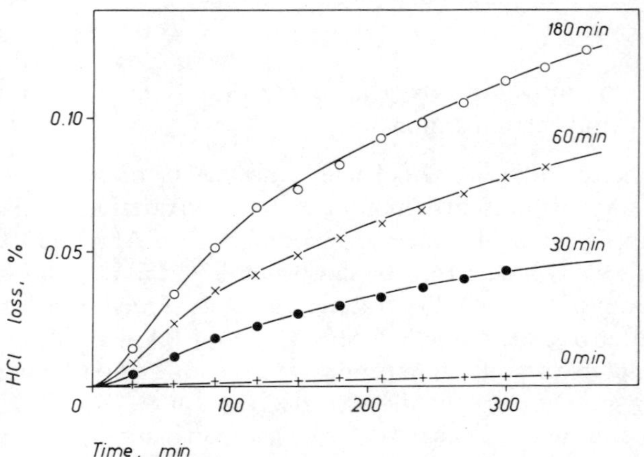

FIG. 5. Loss of HCl during oxidation of virgin and predegraded PVC powder samples. Experimental conditions: PVC powder predegraded at 180 °C in Ar; predegradation times are given on each curve; oxidised at 110 °C in 1 atm O_2.

The rate of HCl evolution increases with increasing polyene content and oxygen partial pressure. Its temperature dependence is characterised by an activation energy of 12 kcal/mol. Due to the different activation energies of the participating chemical processes the kinetic picture changes markedly with increasing temperature. At higher temperatures, e.g., at 145 °C (Fig. 6) the HCl evolution rate of the virgin sample in oxygen at first gradually

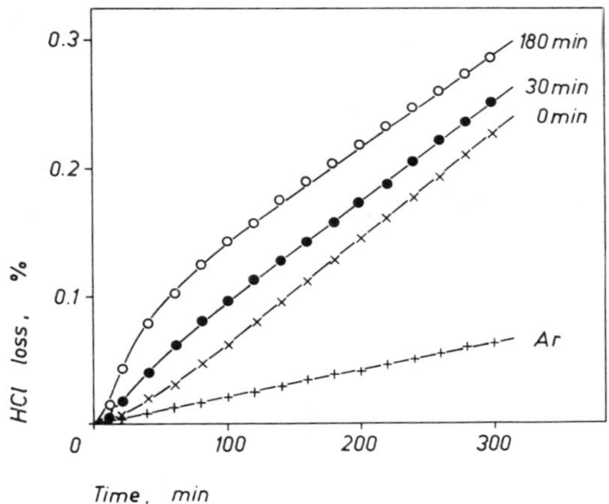

FIG. 6. Loss of HCl during oxidation of virgin and predegraded PVC powder samples at 145 °C in 1 atm O_2. PVC predegraded as in Fig. 5.

increases, then reaches a steady-state. The same sample in an inert atmosphere exhibits a constant rate, much lower than in oxygen. Owing to polyene oxidation, the pre-degraded samples in oxygen have higher rates in the transitory period, but steady-state rates are approximately the same or even somewhat lower than those found for virgin PVC. After this initial period the concentrations of the most important intermediates of thermo-oxidative PVC degradation—the polyenes and the peroxides—reach steady-state values.

On further increase of the temperature of thermo-oxidative PVC degradation, the length of the transitory period decreases and becomes comparable to the time required for thermal equilibrium. Therefore in the literature, the rate of steady-state HCl loss is usually considered as the initial rate.[22] Except for the very early stages of the reaction, the rate of oxygen consumption is proportional to the rate of HCl loss. Due to irregularities and/or impurities in the polymer which do not initiate thermal

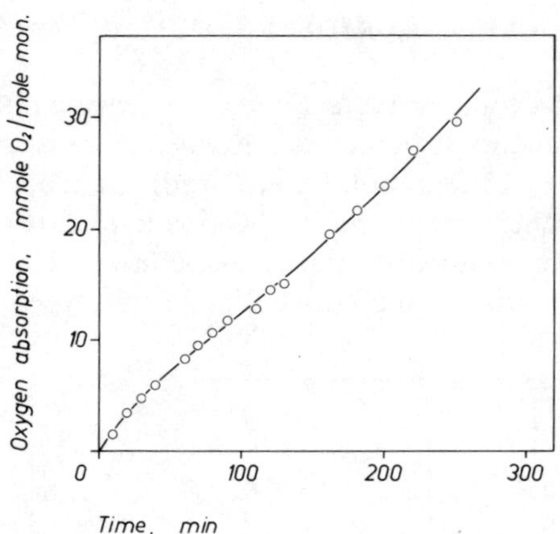

FIG. 7. The amount of oxygen absorbed during thermo-oxidative degradation of PVC at 180 °C, 1 atm O_2.

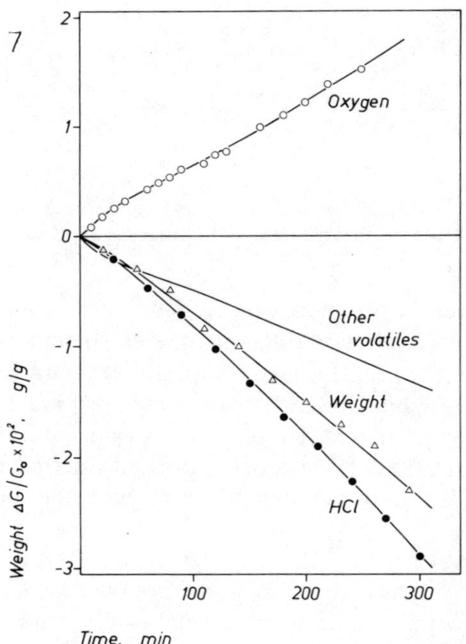

FIG. 8. The weights of absorbed oxygen, eliminated HCl, and polymeric residue and the calculated weight of volatiles (except HCl) during thermo-oxidative degradation of PVC. $\Delta G/G_0$ is the change in weight, related to the original weight of the polymer. Experimental conditions as in Fig. 7.

dehydrochlorination, a small amount of oxygen (approximately 10^{-4} mol O_2/mol monomer unit) is rapidly absorbed in the initial period, as shown in Fig. 7.

If the total amount of oxygen absorbed were bound to the polymer, this would largely compensate for the weight loss caused by HCl loss (Fig. 8). The experiments in pure oxygen, however, show higher weight loss than expected. This indicates that in addition to HCl, other volatile products are formed, some of which also contain oxygen (e.g. water). The amount of these products was calculated from the weight of absorbed oxygen, eliminated HCl and measured weight. Guyot and Benevise also reported the formation of other volatiles during thermo-oxidative degradation in air, especially at advanced stages of HCl loss.[24]

The u.v.-visible spectrum of PVC degraded under thermo-oxidative conditions is basically different from the spectrum of thermally degraded samples (see for example references 20, 21, 36) but similar to the spectrum of oxidised polyenes (Fig. 9).

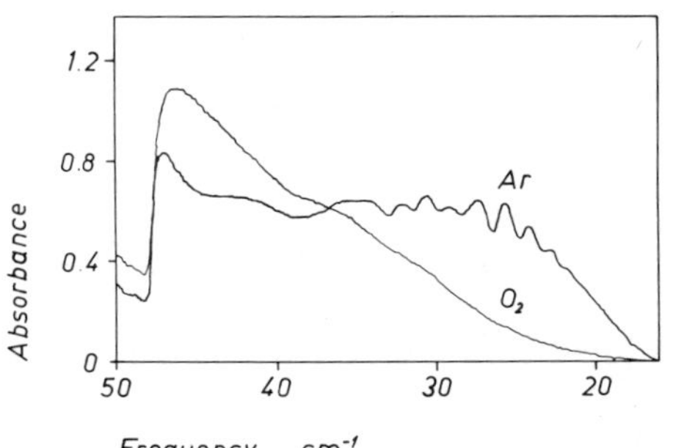

FIG. 9. Visible spectra of degraded PVC. Degradation of powder samples at 200 °C to 0·6% HCl loss in 1 atm of Ar and O_2. Photometry in THF solution (9 g/litre).

The thermo-oxidative degradation of PVC is a complicated chemical process even in the initial stages (up to a few percent of HCl loss). The mechanism is not known in detail, but the main reaction routes, which involve consecutive, competitive and 'feedback' steps are clear. At the temperatures usually applied (above 160 °C) the main reaction routes are as follows:[36,37]

(1) Primary HCl loss and polyene formation take place in a similar process in oxygen and in inert atmospheres.

(2) The polyene sequences are readily oxidised in radical chain reactions.
(3) The peroxides formed decompose rapidly.
(4) The radicals formed in the course of oxidation attack intact monomer units and initiate further HCl loss.

These basic steps are represented in Fig. 10.

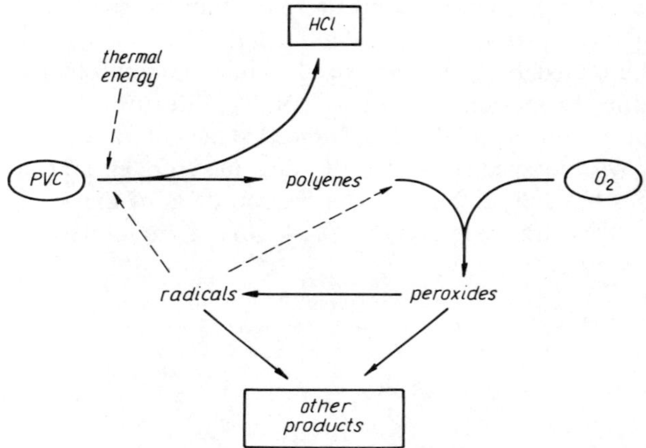

FIG. 10. Reaction scheme for thermo-oxidative degradation of PVC. Dashed arrows indicate initiation.

This mechanism is supported by many experimental data, specially the oxidation kinetics of polyenes and thermo-oxidative degradation in the presence of antioxidants.[38,40] The overall activation energy of HCl loss in the steady-state period of thermo-oxidative degradation is around 30 kcal/mol, practically the same as in the thermal degradation.[36,71] Earlier authors found a lower activation energy for the thermo-oxidative process, but usually at a different stage of the reaction (for example, see reference 13). In terms of the above mechanism one would expect similar activation energies for the two processes because primary initiation occurs in the same way and the kinetic chain length is not expected to change much with temperature.

The increase in rate caused by oxygen may vary somewhat with samples of different origin; it was found to be nearly proportional to the square root of oxygen partial pressure[12,13] (other relations have also been suggested[22]). The ratio of HCl loss rate at 1 atm oxygen to the rate in an inert atmosphere is usually in the range 2–5.

5. THERMO-OXIDATIVE DEGRADATION OF PVC AT HIGH CONVERSIONS

As mentioned in the introduction, thermo-oxidative HCl loss is auto-accelerating (for example, see references 11, 13, 14). The kinetics of HCl loss is accurately described by the following formal autocatalytic equation:

$$d\xi/dt = (k_s + k\xi)(1 - \xi/\xi_\infty) \tag{4}$$

here ξ is the conversion measured by HCl loss, t is the time, k_s the steady-state rate (extrapolated to the beginning of the degradation), and k and ξ_∞

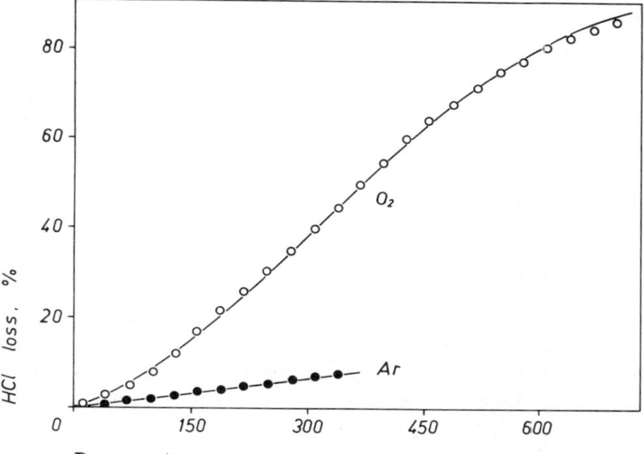

FIG. 11. Loss of HCl from PVC powder degraded at 200 °C in 1 atm Ar and O_2. (The thermo-oxidative curve was calculated according to eqn. (4), using $k_s = 7{\cdot}3 \times 10^{-4}\,\mathrm{min}^{-1}$ and $k = 5{\cdot}15 \times 10^{-3}\,\mathrm{min}^{-1}$.)

are formal constants. High conversion experiments are usually carried out above 200 °C,[13,53] where the transition period is negligibly short, so the steady-state rate is taken as the initial rate. Appreciable deviation is observed only at conversions above 70 % HCl loss if $\xi_\infty = 1$ is used (Fig. 11). The equation used by Talamini,[13] corresponding to $k_s = 0$ and $\xi_\infty = 1$, gives significant deviation at both small and large conversions, but a good description in the middle range.

The steady-state rate gives the same Arrhenius plot as the lower temperature measurements e.g.:

$$k_s = 5{\cdot}37 \times 10^{10} \exp[(-30\,200\,\mathrm{cal/mol})/RT](\mathrm{min}^{-1}) \tag{5}$$

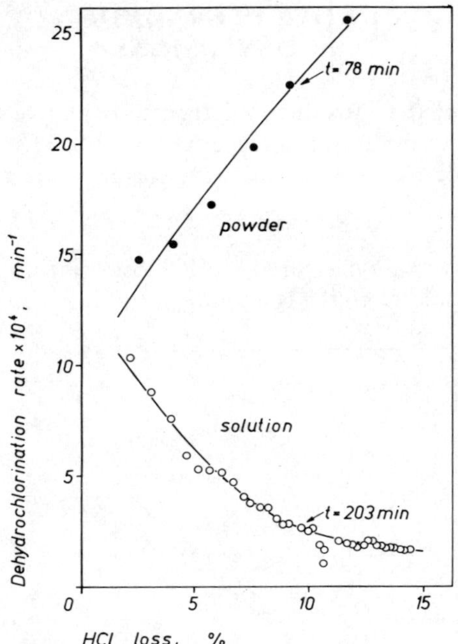

FIG. 12. HCl loss rate as a function of HCl loss from PVC powder and 1,2,4-trichlorobenzene solution (10 g/litre) in oxygen at 210 °C. Degradation time required for 10 % HCl loss is indicated.

whereas k does not give a linear Arrhenius plot, illustrating the complex nature of this parameter.

The auto-accelerating behaviour of the thermo-oxidative degradation of PVC is often regarded as an inherent property.[11,19] The mechanism outlined above does not explain accelerating HCl loss to high conversions but only for the very early stages until the whole process reaches a steady-state. On the other hand auto-acceleration may be the result of HCl catalysis. Except in some early publications[8,9] no doubts have been raised concerning HCl catalysis under oxidative conditions.[1] The chances of effective removal of HCl are better in solution than in PVC powder, so experiments were carried out also in 1,2,4-trichlorobenzene solution. Contrary to the solid-phase thermo-oxidative degradation the rate of HCl loss in solution decreases with conversion (Fig. 12).

A similar although smaller rate decrease takes place in the thermal degradation of PVC solutions.[48] The thermo-oxidative rate remains higher than the thermal rate, however, even at conversions as high as 15%.

This shows that auto-acceleration (except for the transition period) is not an inherent property of thermo-oxidative PVC degradation, but is associated only with solid-phase degradation.

6. CROSS-LINKING AND CHAIN SCISSION IN THERMO-OXIDATIVE DEGRADATION

It has long been known that chain scission and cross-linking occur simultaneously during thermo-oxidative degradation of PVC (for example, see reference 9), but only a few quantitative kinetic measurements have been made.[28,29] Following the kinetics of gel formation is a relatively easy way to determine rates of cross-linking and chain scission if the initial polymer has a random molecular weight distribution and the rates of cross-linking and chain scission are nearly constant. In this case Charlesby–Pinner plots (Fig. 13) (eqn. (6)) give straight lines:[72]

$$s + \sqrt{s} = \frac{p}{q} + \frac{1}{qP_n^\circ} \tag{6}$$

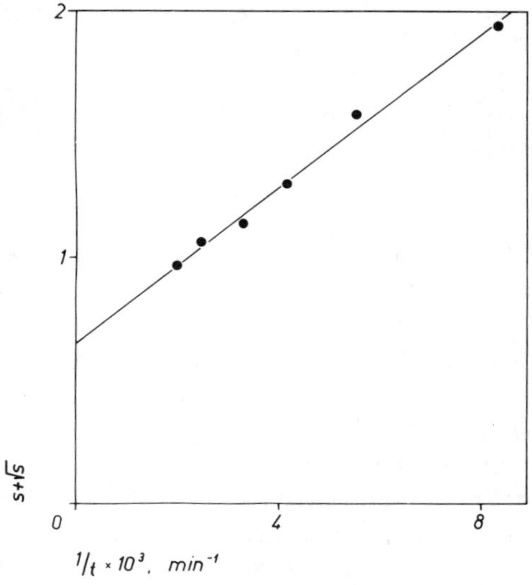

FIG. 13. Charlesby–Pinner plot for PVC powder, degraded in oxygen at 180 °C (s denotes the soluble fraction, t the time).

In eqn. (6), p is the number of scissions per monomer unit; q the proportion of monomer units participating in cross-links; s is the weight fraction of soluble polymer, and P_n° is the number average degree of polymerisation of the original polymer.

According to eqn. (6) the intercept of the Charlesby–Pinner plot gives the ratio of the rate of chain scission to cross-linking, and the rate of cross-linking may be calculated from the slope.

Virgin PVC usually has a molecular weight distribution close to random. Linear Charlesby–Pinner plots (e.g., Fig. 13) were obtained in the thermo-oxidative degradation of several different PVC samples.[73] At 180 °C in pure oxygen the rate of scission varied in the range $1{\cdot}4 \times 10^{-6}$–13×10^{-6} and the rate of cross-linking in the range 3×10^{-6}–20×10^{-6} min^{-1}. Their ratio was found to lie between 0·4 and 0·7.

It should be noted that there is no correlation between the scission rate measured and the amount of weight loss mentioned earlier: the amount of volatile products formed in addition to HCl was much higher than could be reasonably formed in the course of chain scission. The explanation may be that (similar to the formation of benzene during thermal degradation where no chain scission is observed[23]) the majority of volatile products break away from the polymer without any scission of the main chain in a rearrangement process.

As in thermally degraded PVC[59] a considerable, although smaller portion of cross-links were reopened by maleic anhydride treatment of thermo-oxidatively degraded samples.[73] This seems to indicate that reversible Diels–Alder type reactions are also involved in cross-linking under thermo-oxidative conditions.

7. CONCLUDING REMARKS AND OPEN PROBLEMS

Although the main features of thermo-oxidative degradation of PVC are relatively well understood, some important questions remain unanswered. It was demonstrated that thermally initiated HCl loss, followed by oxidation of polyene sequences and further HCl loss initiated by radicals, play a decisive role in the process. The primary HCl loss process is similar in pure thermal and thermo-oxidative degradation. This is interesting from the point of view of stabilisation. The most important aspect—also under thermo-oxidative conditions—is to prevent the unzipping of HCl in the primary process, as the normal PVC chain is much less sensitive to oxidation than the polyene sequences. This is reflected in the practice of

PVC processing, since the heat stabilisers used under non-oxidative and oxidative conditions are the same chemical compounds. The use of antioxidants seems to be controversial because although they retard HCl loss, they also prevent or slow down the oxidation of polyenes so that the development of colour may indicate destabilisation. Despite the great efforts of several laboratories throughout the world, many details of PVC degradation remain unclear. There are a great number of open questions especially in the field of thermo-oxidative degradation. Only a few of these will be raised here.

Is oxygen capable of reaction with unterminated polyene sequences, i.e., does oxidation compete with fast unzipping and termination of primary HCl loss?

What is the role of oxidation-sensitive weak sites (originally present in the polymer) in the early stages of degradation?

What are the relative rates and the relative importance of the main participating processes?

What are the mechanisms of chain scission, cross-linking and formation of volatile products?

What quantitative relationships exist between the extent of thermo-oxidative degradation and the deterioration of the mechanical properties of the polymer?

An answer to these and similar questions requires much further research in the field of thermo-oxidative degradation of PVC.

ACKNOWLEDGEMENTS

The authors wish to express thanks to B. Turcsányi and B. Iván for their helpful assistance and valuable discussions. Thanks are also due to E. Gerber and K. Pula for technical assistance.

REFERENCES

1. MAYER, Z., *J. Macromol. Sci., Rev. Macromol. Chem.*, **C10** (1974) p. 263.
2. AYREY, G., HEAD, B. C. and POLLER, R. C., *J. Polym. Sci., Macromol. Rev.*, **8** (1974) p. 1.
3. BRAUN, D., In *Degradation and Stabilization of Polymers*, G. Geuskens, Ed., New York, John Wiley, 1975, p. 23.

4. DAVID, C., In *Comprehensive Chemical Kinetics*, Vol 14, C. H. Bamford and C. F. H. Tipper, Eds., Amsterdam, Elsevier, 1975, p. 78.
5. NASS, L. I., In *Encyclopedia of PVC*, Vol 1, L. I. Nass, Ed., New York, Dekker, 1976, p. 271.
6. TROITSKII, B. B. and TROITSKAYA, L. S., *Vysokomol. Soedin.*, **A20,** (1978) p. 1443.
7. PUDOV, V. S., *Plast. Massy* (2) (1976) p. 18.
8. ARLMAN, E. J., *J. Polym. Sci.*, **12** (1954) p. 543.
9. ARLMAN, E. J., *J. Polym. Sci.*, **12** (1954) p. 547.
10. BRAUN, D. and BENDER, R. F., *Eur. Polym. J.—Supplement*, (1969) p. 269.
11. GEDDES, W. C., *Eur. Polym. J.*, **3** (1967) p. 267.
12. MINSKER, K. S., BERLIN, AL. AL. and ABDULLIN, M. N., *Vysokomol. Soedin.*, **B16** (1974) p. 439.
13. TALAMINI, G. and PEZZIN, G., *Makromol. Chem.*, **39** (1960) p. 26.
14. JASCHING, W., *Kunststoffe*, **52** (1962) p. 458.
15. GUPTA, S. N , KENNEDY, J. P., NAGY, T. T , TÜDŐS, F. and KELEN, T., *J. Macromol. Sci.-Chem.*, in press.
16. LUTHER, H. and KRÜGER, H., *Kunststoffe*, **56** (1966) p. 74.
17. GUYOT, A., BENEVISE, J. P. and TRAMBOUZE, Y., *J. Appl. Polym. Sci.*, **6** (1962) p. 103.
18. DANFORTH, J. D. and TAKEUCHI, T., *J. Polym. Sci., Polym. Chem. Ed.*, **11** (1973) p. 2091.
19. ONOZUKA, M. and ASAHINA, M., *J. Macromol. Sci. Rev. Macromol. Chem.*, **C3** (1969) p. 235.
20. PALMA, G. and CARENZA, M., *J. Appl. Polym. Sci.*, **16** (1972) p. 2485.
21. BRAUN, D., *Gummi, Asbest, Kunststoffe*, **24** (1971) p. 1116.
22. ABBAS, K. B. and SÖRVIK, E. M., *J. Appl. Polym. Sci.*, **17** (1973) p. 3577.
23. NAGY, T. T., TURCSÁNYI, B., KELEN, T. and TÜDŐS, F., *Reaction Kinetics and Catalysis Lett.*, 1976, **5** (1976) p. 309.
24. GUYOT, A. and BENEVISE, J. P., *J. Appl. Polym. Sci.*, **6** (1962) p. 489.
25. WOLKÓBER, Z., *Angew. Makromol. Chem.*, **3** (1968) p. 38.
26. REICHERT, W., WOLKÓBER, Z. and KRAUSE, H., *Plaste u. Kaut.*, **13** (1966) p. 454.
27. KRAUSE, H., REICHERT, W. and THINIUS, K., *Plaste u. Kaut.*, **18** (1971) p. 194.
28. BENGOUGH, W. I. and SHARPE, H. M., *Makromol. Chem.*, **66** (1963) p. 45.
29. KURZWEIL, K. and KRATOCHVIL, P., *Coll. Czech. Chem. Comm.*, **34** (1969) p. 1429.
30. MORIKAWA, T., *Kagaku to Kogyo*, **38** (1964) p. 42.
31. MORIKAWA, T., *Kagaku to Kogyo*, **39** (1965) p. 325.
32. DRUESEDOW, D. and GIBBS, C. F., *Modern Plastics*, **30** (1953) p. 123.
33. RIECHE, A., GRIMM, A. and MÜCKE, H., *Kunststoffe*, **52** (1962) p. 265.
34. VALKO, L., *J. Polym. Sci. Part C.*, **16** (1967) p. 1979.
35. VALKO, L., *J. Polym. Sci. Part C*, **16** (1967) p. 545.
36. NAGY, T. T., KELEN, T., TURCSÁNYI, B. and TÜDŐS, F., *Angew. Makromol. Chem.*, **66** (1978) p. 193.
37. NAGY, T. T., TURCSÁNYI, B., KELEN, T. and TÜDŐS, F., *Reaction Kinetics and Catalysis Lett.*, **8** (1978) p. 7.
38. MINSKER, K. S. and PAKHOMOVA, I. K., *Vysokomol. Soedin.*, **A11** (1969) p. 646.
39. STAPFER, C. H. and GRANICK, J. D., *J. Polym. Sci. Part A*1, **9** (1971) p. 2625.

40. MINSKER, K. S., KRATS, E. O. and PAKHOMOVA, I. K., *Vysokomol. Soedin.*, **A12** (1970) p. 483.
41. BENGOUGH, W. I. and SHARPE, H. M., *Makromol. Chem.*, **66** (1963) p. 31.
42. GEDDES, W. C., *Eur. Polym. J.*, **3** (1967) p. 733.
43. MINSKER, K. S., ABDULLIN, M. I. and GARIPOVA, G. A., *Dokl. Akad. Nauk SSSR—Ser. Khim.*, **218** (1974) p. 851.
44. ABDULLIN, M. I., MALINSKAYA, V. P., KOLESOV, S. V. and MINSKER, K. S., Second Internat. Symp. on Poly(vinyl-chloride), Lyon-Villeurbanne, July 1976. A. Michel, Ed., Lyon-Villeurbanne, CNRS, Preprints p. 273.
45. MICHEL, A., CASTANEDA, E. and GUYOT, A., *J. Macromol. Sci.-Chem.*, **A12** (1978) p. 227.
46. ABDULLIN, M. I., GATULLIN, R. F., MINSKER, K. S., KEFELI, A. A. and RAZUMOVSKII, S. D., *Eur. Polym. J.*, **14** (1978) p. 811.
47. BRAUN, D. and WOLF, M., *Angew. Makromol. Chem.*, **70** (1978) p. 71.
48. TÜDŐS, F. and KELEN, T., In *Macromolecular Chemistry–8*, K. Saarela, Ed., London, Butterworths, 1973, p. 393.
49. TÜDŐS, F., KELEN, T., NAGY, T. T. and TURCSÁNYI, B., *Pure Appl. Chem.*, **38** (1974) p. 201.
50. CZAKÓ, E., VYMAZAL, Z., VYMAZALOVA, Z. and STEPEK, J., *Eur. Polym. J.*, **13** (1977) p. 847.
51. RABEK, J. F., CANBÄCK, G. and RANBY, B., *J. Appl. Polym. Sci.*, **21** (1977) p. 2211.
52. PACIOREK, K. L., KRATZER, R. H., KAUFMAN, U., NAKAHARA, J. and HARTSTEIN, A. M., *J. Appl. Polym. Sci.*, **18** (1974) p. 3723.
53. PARINET, B., MOST, J. M., JOULAIN, P. and VANTELON, J. P., *J. Chim. Phys.*, **74** (1977) p. 539.
54. RABEK, J. F., In *Comprehensive Chemical Kinetics*, Vol 14, C. H. Bamford and C. F. H. Tipper, Eds., Amsterdam, Elsevier, 1975, p. 522.
55. EMANUEL, N. M., ZAYKOV, G. E. and MAYZUS, Z. K., *Role of Medium in Radical-chain Oxidation of Organic Compounds* (in Russian), Moscow, Izdatelstvo Nauka, 1973.
56. KELEN, T., GALAMBOS, G., TÜDŐS, F. and BÁLINT, G., *Eur. Polym. J.*, **5** (1969) p. 617.
57. KELEN, T., *J. Macromol. Sci.-Chem.*, **A12** (1978) p. 349.
58. KELEN, T., NAGY, T. T. and TÜDŐS, F., *Reaction Kinetics and Catalysis Lett.*, **1** (1974) p. 93.
59. KELEN, T., IVÁN, B., NAGY, T. T., TURCSÁNYI, B., TÜDŐS, F. and KENNEDY J. P., *Polymer Bulletin*, **1** (1978) p. 79.
60. MAYO, F. R., *Accounts Chem. Res.*, **1** (1968) p. 193.
61. GAGARINA, A. B., KASAIKINA, O. T. and EMANUEL, N. M., *Dokl. Akad. Nauk SSSR—Ser. Khim.*, **195** (1970) p. 387.
62. CHOU, H. E. and BREENE, W. M., *J. Food Sci.*, **37** (1972) p. 66.
63. MATSUMOTO, T., MUNE, I. and WATATANI, S., *J. Polym. Sci. Part A1*, **7** (1969) p. 1609.
64. DANIELS, V. D. and REES, N. H., *J. Polym. Sci., Polym. Chem. Ed.*, **12** (1974) p. 2115.
65. NAGY, T. T., KELEN, T., TURCSÁNYI, B. and TÜDŐS, F., *J. Polym. Sci., Polym. Chem. Ed.*, **15** (1977) p. 853.

66. Foote, Ch. S., *Pure Appl. Chem.*, **27** (1971) p. 635.
67. Nagy, T. T., Kelen, T., Turcsányi, B. and Tüdős, F., *Plaste u. Kaut.*, **23** (1976) p. 894.
68. Nagy, T. T., Kelen, T., Turcsányi, B. and Tüdős, F., *Reaction Kinetics and Catalysis Lett.*, **5** (1976) p. 303.
69. Mair, R. D. and Graupner, A. J., *Anal. Chem.*, **36** (1964) p. 194.
70. Guyot, A. and Bert, M., *Polymer Preprints*, **12** (1971) p. 303.
71. Abbas, K. B. and Sörvik, E. M., *J. Appl. Polym. Sci.*, **17** (1973) p. 3567.
72. Charlesby, A. and Pinner, S. H., *Proc. Roy. Soc.*, **A249** (1959) p. 367.
73. Iván, B., Kelen, T., Turcsányi, B., Nagy, T. T. and Tüdős, F., to be published.

INDEX